D1357645

Springer Monographs in Mathematics

Springer
Berlin
Heidelberg
New York
Barcelona
Hong Kong
London
Milan
Paris
Singapore
Tokyo

Jean-Pierre Serre

Complex Semisimple Lie Algebras

Translated from the French by G. A. Jones

Reprint of the 1987 Edition

 Springer

Jean-Pierre Serre
Collège de France
75231 Paris Cedex 05
France
e-mail: serre@dmi.ens.fr

Translated By:
G. A. Jones
University of Southampton
Faculty of Mathematical Studies
Southampton SO9 5NH
United Kingdom

Library of Congress Cataloging-in-Publication Data applied for

Die Deutsche Bibliothek - CIP-Einheitsaufnahme
Serre, Jean-Pierre:
Complex semisimple Lie algeras / Jean-Pierre Serre. Transl. from the
French by G. A. Jones. - Reprint of the 1987 ed.. - Berlin ;
Heidelberg ; New York ; Barcelona ; Hong Kong ; London ; Milan ; Paris ;
Singapore ; Tokyo : Springer, 2001
 (Springer monographs in mathematics)
 Einheitssacht.: Algèbres de Lie semi-simples complexes <engl.>
 ISBN 3-540-67827-1

This book is a translation of the original French edition *Algèbres de Lie Semi-Simples Complexes*, published by Benjamin, New York in 1966.

Mathematics Subject Classification (2000): 17B05, 17B20

ISSN 1439-7382
ISBN 3-540-67827-1 Springer-Verlag Berlin Heidelberg New York

This work is subject to copyright. All rights are reserved, whether the whole or part of the material is concerned, specifically the rights of translation, reprinting, reuse of illustrations, recitation, broadcasting, reproduction on microfilm or in any other way, and storage in data banks. Duplication of this publication or parts thereof is permitted only under the provisions of the German Copyright Law of September 9, 1965, in its current version, and permission for use must always be obtained from Springer-Verlag. Violations are liable for prosecution under the German Copyright Law.

Springer-Verlag Berlin Heidelberg New York
a member of BertelsmannSpringer Science+Business Media GmbH

© Springer-Verlag Berlin Heidelberg 2001
Printed in Germany

The use of general descriptive names, registered names, trademarks, etc. in this publica-tion does not imply, even in the absence of a specific statement, that such names are exempt from the relevant protective laws and regulations and therefore free for general use.

Typeset by Asco Trade Typesetting Ltd., Hong Kong

Printed on acid-free paper   SPIN 10734431   41/3142LK – 5 4 3 2 1 0

Jean-Pierre Serre

Complex Semisimple Lie Algebras

Translated from the French by G. A. Jones

Springer-Verlag
New York Berlin Heidelberg
London Paris Tokyo

Jean-Pierre Serre
Collège de France
75231 Paris Cedex 05
France

Translated by:
G. A. Jones
University of Southampton
Faculty of Mathematical Studies
Southampton SO9 5NH
United Kingdom

AMS Classifications: 17B05, 17B20

With 6 Illustrations

Library of Congress Cataloging-in-Publication Data
Serre, Jean-Pierre.
  Complex semisimple Lie algebras.
  Translation of: Algèbres de Lie semi-simples complexes.
  Bibliography: p.
  Includes index.
  1. Lie algebras.   I. Title.
QA251.S4713   1987   512′.55   87-13037

This book is a translation of the original French edition, *Algèbres de Lie Semi-Simples Complexes*.
© 1966 by Benjamin, New York.

© 1987 by Springer-Verlag New York Inc.
All rights reserved. This work may not be translated or copied in whole or in part without the
written permission of the publisher (Springer-Verlag, 175 Fifth Avenue, New York, New York
10010, USA), except for brief excerpts in connection with reviews or scholarly analysis. Use in
connection with any form of information storage and retrieval, electronic adaptation, computer
software, or by similar or dissimilar methodology now known or hereafter developed is forbidden.
The use of general descriptive names, trade names, trademarks, etc. in this publication, even if
the former are not especially identified, is not to be taken as a sign that such names, as understood
by the Trade Marks and Merchandise Marks Act, may accordingly be used freely by anyone.

Typeset by Asco Trade Typesetting Ltd., Hong Kong.
Printed and bound by R. R. Donnelley and Sons, Harrisonburg, Virginia.
Printed in the United States of America.

9 8 7 6 5 4 3 2 1

ISBN 0-387-96569-6 Springer-Verlag New York Berlin Heidelberg
ISBN 3-540-96569-6 Springer-Verlag Berlin Heidelberg New York

Preface

These notes are a record of a course given in Algiers from 10th to 21st May, 1965. Their contents are as follows.

The first two chapters are a summary, without proofs, of the general properties of nilpotent, solvable, and semisimple Lie algebras. These are well-known results, for which the reader can refer to, for example, Chapter I of Bourbaki or my Harvard notes.

The theory of complex semisimple algebras occupies Chapters III and IV. The proofs of the main theorems are essentially complete; however, I have also found it useful to mention some complementary results without proof. These are indicated by an asterisk, and the proofs can be found in Bourbaki, *Groupes et Algèbres de Lie*, Paris, Hermann, 1960–1975, Chapters IV–VIII.

A final chapter shows, without proof, how to pass from Lie algebras to Lie groups (complex—and also compact). It is just an introduction, aimed at guiding the reader towards the topology of Lie groups and the theory of algebraic groups.

I am happy to thank MM. Pierre Gigord and Daniel Lehmann, who wrote up a first draft of these notes, and also Mlle. Françoise Pécha who was responsible for the typing of the manuscript.

Jean-Pierre Serre

Contents

CHAPTER I

Nilpotent Lie Algebras and
Solvable Lie Algebras

The Lie algebras considered in this chapter are finite-dimensional algebras over a field k. In Secs. 7 and 8 we assume that k has characteristic 0. The Lie bracket of x and y is denoted by $[x, y]$, and the map $y \mapsto [x, y]$ by ad x.

1. Lower Central Series

Let $\mathfrak{g}$ be a Lie algebra. The *lower central series* of $\mathfrak{g}$ is the descending series $(C^n \mathfrak{g})_{n \geqslant 1}$ of ideals of $\mathfrak{g}$ defined by the formulae

$$C^1 \mathfrak{g} = \mathfrak{g}$$
$$C^n \mathfrak{g} = [\mathfrak{g}, C^{n-1} \mathfrak{g}] \quad \text{if } n \geqslant 2.$$

We have

$$C^2 \mathfrak{g} = [\mathfrak{g}, \mathfrak{g}]$$

and

$$[C^n \mathfrak{g}, C^m \mathfrak{g}] \subset C^{n+m} \mathfrak{g}.$$

2. Definition of Nilpotent Lie Algebras

Definition 1. *A Lie algebra $\mathfrak{g}$ is said to be nilpotent if there exists an integer n such that $C^n \mathfrak{g} = 0$.*

More precisely, one says that $\mathfrak{g}$ is nilpotent of class $\leqslant r$ if $C^{r+1} \mathfrak{g} = 0$. For $r = 1$, this means that $[\mathfrak{g}, \mathfrak{g}] = 0$; that is, $\mathfrak{g}$ is abelian.

Proposition 1. *The following conditions are equivalent:*

(i) $\mathfrak{g}$ *is nilpotent of class* $\leqslant r$.

(ii) *For all* $x_0, \ldots, x_r \in \mathfrak{g}$, *we have*

$$[x_0, [x_1, [\ldots, x_r]\ldots]] = (\operatorname{ad} x_0)(\operatorname{ad} x_1)\ldots(\operatorname{ad} x_{r-1})(x_r) = 0.$$

(iii) *There is a descending series of ideals*

$$\mathfrak{g} = \mathfrak{a}_0 \supset \mathfrak{a}_1 \supset \cdots \supset \mathfrak{a}_r = 0$$

such that $[\mathfrak{g}, \mathfrak{a}_i] \subset \mathfrak{a}_{i+1}$ *for* $0 \leqslant i \leqslant r - 1$.

Now recall that the *center* of a Lie algebra $\mathfrak{g}$ is the set of $x \in \mathfrak{g}$ such that $[x, y] = 0$ for all $y \in \mathfrak{g}$. It is an abelian ideal of $\mathfrak{g}$.

Proposition 2. *Let* $\mathfrak{g}$ *be a Lie algebra and let* $\mathfrak{a}$ *be an ideal contained in the center of* $\mathfrak{g}$. *Then:*

$$\mathfrak{g} \text{ is nilpotent} \Leftrightarrow \mathfrak{g}/\mathfrak{a} \text{ is nilpotent.}$$

The above two propositions show that the nilpotent Lie algebras are those one can form from abelian algebras by successive "central extensions."

(Warning: an extension of nilpotent Lie algebras is not in general nilpotent.)

3. An Example of a Nilpotent Algebra

Let V be a vector space of finite dimension n. A *flag* $D = (D_i)$ of V is a descending series of vector subspaces

$$V = D_0 \supset D_1 \supset \cdots \supset D_n = 0$$

of V such that codim $D_i = i$.

Let D be a flag, and let $\mathfrak{n}(D)$ be the Lie subalgebra of $\operatorname{End}(V) = \mathfrak{gl}(V)$ consisting of the elements x such that $x(D_i) \subset D_{i+1}$. One can verify that $\mathfrak{n}(D)$ is a nilpotent Lie algebra of class $n - 1$.

4. Engel's Theorems

Theorem 1. *For a Lie algebra* $\mathfrak{g}$ *to be nilpotent, it is necessary and sufficient for* ad x *to be nilpotent for each* $x \in \mathfrak{g}$.

(This condition is clearly necessary, cf. Proposition 1.)

Theorem 2. *Let* V *be a finite-dimensional vector space and* $\mathfrak{g}$ *a Lie subalgebra of* $\operatorname{End}(V)$ *consisting of nilpotent endomorphisms. Then:*

(a) $\mathfrak{g}$ *is a nilpotent Lie algebra.*
(b) *There is a flag D of V such that* $\mathfrak{g} \subset \mathfrak{n}(D)$.

We can reformulate the above theorem in terms of $\mathfrak{g}$-modules. To do this, we recall that if $\mathfrak{g}$ is a Lie algebra and V a vector space, then a Lie algebra homomorphism $\phi: \mathfrak{g} \to \text{End}(V)$ is called a $\mathfrak{g}$-module structure on V; one also says that ϕ is a linear representation of $\mathfrak{g}$ on V. An element $v \in V$ is called invariant under $\mathfrak{g}$ (for the given $\mathfrak{g}$-module structure) if $\phi(x)v = 0$ for all $x \in \mathfrak{g}$. (This surprising terminology arises from the fact that, if $k = \mathbf{R}$ or $\mathbf{C}$, and if ϕ is associated with a representation of a connected Lie group G on V, then v is invariant under $\mathfrak{g}$ if and only if it is invariant—this time in the usual sense—under G.)

With this terminology, Theorem 2 gives:

Theorem 2'. *Let $\phi: \mathfrak{g} \to \text{End}(V)$ be a linear representation of a Lie algebra $\mathfrak{g}$ on a nonzero finite-dimensional vector space V. Suppose that $\phi(x)$ is nilpotent for all $x \in \mathfrak{g}$. Then there exists an element $v \neq 0$ of V which is invariant under $\mathfrak{g}$.*

5. Derived Series

Let $\mathfrak{g}$ be a Lie algebra. The *derived series* of $\mathfrak{g}$ is the descending series $(D^n\mathfrak{g})_{n \geqslant 1}$ of ideals of $\mathfrak{g}$ defined by the formulae

$$D^1\mathfrak{g} = \mathfrak{g}$$
$$D^n\mathfrak{g} = [D^{n-1}\mathfrak{g}, D^{n-1}\mathfrak{g}] \qquad \text{if } n \geqslant 2.$$

One usually writes $D\mathfrak{g}$ for $D^2\mathfrak{g} = [\mathfrak{g}, \mathfrak{g}]$.

6. Definition of Solvable Lie Algebras

Definition 2. *A Lie algebra $\mathfrak{g}$ is said to be solvable if there exists an integer n such that $D^n\mathfrak{g} = 0$.*

Here again, one says that $\mathfrak{g}$ is solvable of derived length $\leqslant r$ if $D^{r+1}\mathfrak{g} = 0$.

EXAMPLES. 1. Every nilpotent algebra is solvable.

2. Every subalgebra, every quotient, and every extension of solvable algebras is solvable.

3. Let $D = (D_i)$ be a flag of a vector space V, and let $\mathfrak{b}(D)$ be the subalgebra of $\text{End}(V)$ consisting of the $x \in \text{End}(V)$ such that $x(D_i) \subset D_i$ for all i. The algebra $\mathfrak{b}(D)$ (a "Borel algebra") is solvable.

Proposition 3. *The following conditions are equivalent*:

(i) $\mathfrak{g}$ *is solvable of derived length* $\leqslant r$.

(ii) *There is a descending series of ideals of* $\mathfrak{g}$:

$$\mathfrak{g} = \mathfrak{a}_0 \supset \mathfrak{a}_1 \supset \cdots \supset \mathfrak{a}_r = 0$$

such that $[\mathfrak{a}_i, \mathfrak{a}_i] \subset \mathfrak{a}_{i+1}$ *for* $0 \leqslant i \leqslant r - 1$ *(which amounts to saying that that the quotients* $\mathfrak{a}_i/\mathfrak{a}_{i+1}$ *are abelian).*

Thus one can say that solvable Lie algebras are those obtained from abelian Lie algebras by successive "extensions" (not necessarily central).

7. Lie's Theorem

We assume that k is algebraically closed (and of characteristic zero).

Theorem 3. *Let* $\phi: \mathfrak{g} \to \mathrm{End}(V)$ *be a finite-dimensional linear representation of a Lie algebra* $\mathfrak{g}$. *If* $\mathfrak{g}$ *is solvable, there is a flag* D *of* V *such that* $\phi(\mathfrak{g}) \subset \mathfrak{b}(D)$.

This theorem can be rephrased in the following equivalent forms.

Theorem 3'. *If* $\mathfrak{g}$ *is solvable, the only finite-dimensional* $\mathfrak{g}$-*modules which are simple* (irreducible in the language of representation theory) *are one dimensional.*

Theorem 3''. *Under the hypotheses of Theorem 3, if* $V \neq 0$ *there exists an element* $v \neq 0$ *of* V *which is an eigenvector for every* $\phi(x)$, $x \in \mathfrak{g}$.

The proof of these theorems uses the following lemma.

Lemma. *Let* $\mathfrak{g}$ *be a Lie algebra,* $\mathfrak{h}$ *an ideal of* $\mathfrak{g}$, *and* $\phi: \mathfrak{g} \to \mathrm{End}(V)$ *a finite-dimensional linear representation of* $\mathfrak{g}$. *Let* v *be a nonzero element of* V *and let* λ *be a linear form on* $\mathfrak{h}$ *such that* $\lambda(h)v = \phi(h)v$ *for all* $h \in \mathfrak{h}$. *Then* λ *vanishes on* $[\mathfrak{g}, \mathfrak{h}]$.

8. Cartan's Criterion

It is as follows:

Theorem 4. *Let* V *be a finite-dimensional vector space and* $\mathfrak{g}$ *a Lie subalgebra of* $\mathrm{End}(V)$. *Then*:

$$\mathfrak{g} \text{ is solvable} \Leftrightarrow \mathrm{Tr}(x \circ y) = 0 \qquad \text{for all } x \in \mathfrak{g}, \, y \in [\mathfrak{g}, \mathfrak{g}].$$

(This implication $\Rightarrow$ is an easy corollary of Lie's theorem.)

CHAPTER II

Semisimple Lie Algebras
(General Theorems)

In this chapter, the base field k is a field of characteristic zero. The Lie algebras and vector spaces considered have finite dimension over k.

1. Radical and Semisimplicity

Let $\mathfrak{g}$ be a Lie algebra. If $\mathfrak{a}$ and $\mathfrak{b}$ are solvable ideals of $\mathfrak{g}$, the ideal $\mathfrak{a} + \mathfrak{b}$ is also solvable, being an extension of $\mathfrak{b}/(\mathfrak{a} \cap \mathfrak{b})$ by $\mathfrak{a}$. Hence there is a largest solvable ideal $\mathfrak{r}$ of $\mathfrak{g}$. It is called the *radical* of $\mathfrak{g}$.

Definition 1. *One says that $\mathfrak{g}$ is semisimple if its radical $\mathfrak{r}$ is 0.*

This amounts to saying that $\mathfrak{g}$ has no abelian ideals other than 0.

EXAMPLE. If V is a vector space, the subalgebra $\mathfrak{sl}(V)$ of $\mathrm{End}(V)$ consisting of the elements of trace zero is semisimple.

(See Sec. 7 for more examples.)

Theorem 1. *Let $\mathfrak{g}$ be a Lie algebra and $\mathfrak{r}$ its radical.*

(a) $\mathfrak{g}/\mathfrak{r}$ *is semisimple.*
(b) *There is a Lie subalgebra $\mathfrak{s}$ of $\mathfrak{g}$ which is a complement for $\mathfrak{r}$.*

If $\mathfrak{s}$ satisfies the condition in (b), the projection $\mathfrak{s} \to \mathfrak{g}/\mathfrak{r}$ is an isomorphism, showing (with the aid of (a)) that $\mathfrak{s}$ is semisimple. Thus $\mathfrak{g}$ is a *semidirect product* of a semisimple algebra and a solvable ideal (a "Levi decomposition").

2. The Cartan–Killing Criterion

Let $\mathfrak{g}$ be a Lie algebra. A bilinear form $B\colon \mathfrak{g} \times \mathfrak{g} \to k$ on $\mathfrak{g}$ is said to be invariant if we have

$$B([x, y], z) + B(y, [x, z]) = 0 \qquad \text{for all } x, y, z \in \mathfrak{g}.$$

The Killing form $B(x, y) = \mathrm{Tr}(\mathrm{ad}\, x \circ \mathrm{ad}\, y)$ is invariant and symmetric.

Lemma. *Let B be an invariant bilinear form on $\mathfrak{g}$, and $\mathfrak{a}$ an ideal of $\mathfrak{g}$. Then the orthogonal space $\mathfrak{a}'$ of $\mathfrak{a}$ with respect to B is an ideal of $\mathfrak{g}$.*

(By definition, $\mathfrak{a}'$ is the set of all $y \in \mathfrak{g}$ such that $B(x, y) = 0$ for all $x \in \mathfrak{a}$.)

Theorem 2 (Cartan–Killing Criterion). *A Lie algebra is semisimple if and only if its Killing form is nondegenerate.*

3. Decomposition of Semisimple Lie Algebras

Theorem 3. *Let $\mathfrak{g}$ be a semisimple Lie algebra, and $\mathfrak{a}$ an ideal of $\mathfrak{g}$. The orthogonal space $\mathfrak{a}'$ of $\mathfrak{a}$, with respect to the Killing form of $\mathfrak{g}$, is a complement for $\mathfrak{a}$ in $\mathfrak{g}$; the Lie algebra $\mathfrak{g}$ is canonically isomorphic to the product $\mathfrak{a} \times \mathfrak{a}'$.*

Corollary. *Every ideal, every quotient, and every product of semisimple algebras is semisimple.*

Definition 2. *A Lie algebra $\mathfrak{s}$ is said to be simple if:*

(a) *it is not abelian,*
(b) *its only ideals are 0 and $\mathfrak{s}$.*

EXAMPLE. The algebra $\mathfrak{sl}(V)$ is simple provided that $\dim V \geqslant 2$.

Theorem 4. *A Lie algebra $\mathfrak{g}$ is semisimple if and only if it is isomorphic to a product of simple algebras.*

In fact, this decomposition is unique. More precisely:

Theorem 4'. *Let $\mathfrak{g}$ be a semisimple Lie algebra, and $(\mathfrak{a}_i)$ its minimal nonzero ideals. The ideals $\mathfrak{a}_i$ are simple Lie algebras, and $\mathfrak{g}$ can be identified with their product.*

Clearly, if $\mathfrak{s}$ is simple we have $\mathfrak{s} = [\mathfrak{s}, \mathfrak{s}]$. Thus Theorem 4 implies:

Corollary. *If $\mathfrak{g}$ is semisimple then $\mathfrak{g} = [\mathfrak{g}, \mathfrak{g}]$.*

4. Derivations of Semisimple Lie Algebras

First recall that if A is an algebra, a *derivation* of A is a linear mapping $D: A \to A$ satisfying the identity

$$D(x \cdot y) = Dx \cdot y + x \cdot Dy.$$

The derivations form a Lie subalgebra $\mathrm{Der}(A)$ of $\mathrm{End}(A)$. In particular, this applies to the case where we take A to be a Lie algebra $\mathfrak{g}$. A derivation D of $\mathfrak{g}$ is called *inner* if $D = \mathrm{ad}\, x$ for some $x \in \mathfrak{g}$, or in other words if D belongs to the image of the homomorphism $\mathrm{ad}: \mathfrak{g} \to \mathrm{Der}(\mathfrak{g})$.

Theorem 5. *Every derivation of a semisimple Lie algebra is inner.*

Thus the mapping $\mathrm{ad}: \mathfrak{g} \to \mathrm{Der}(\mathfrak{g})$ is an isomorphism.

Corollary. *Let G be a connected Lie group (real or complex) whose Lie algebra $\mathfrak{g}$ is semisimple. Then the component $\mathrm{Aut}^0\, G$ of the identity in the automorphism group $\mathrm{Aut}\, G$ of G coincides with the inner automorphism group of G.*

This follows from the fact that the Lie algebra of $\mathrm{Aut}^0\, G$ coincides with $\mathrm{Der}(\mathfrak{g})$.

Remark. The automorphisms of $\mathfrak{g}$ induced by the inner automorphisms of G are (by abuse of language) called the *inner automorphisms* of $\mathfrak{g}$. When $\mathfrak{g}$ is semisimple, they form the component of the identity in the group $\mathrm{Aut}(\mathfrak{g})$.

5. Semisimple Elements and Nilpotent Elements

Definition 3. *Let $\mathfrak{g}$ be a semisimple Lie algebra, and let $x \in \mathfrak{g}$.*

(a) *x is said to be nilpotent if the endomorphism $\mathrm{ad}\, x$ of $\mathfrak{g}$ is nilpotent.*
(b) *x is said to be semisimple if $\mathrm{ad}\, x$ is semisimple* (that is, diagonalizable after extending the ground field).

Theorem 6. *If $\mathfrak{g}$ is semisimple, every element x of $\mathfrak{g}$ can be written uniquely in the form $x = s + n$, with n nilpotent, s semisimple, and $[s, n] = 0$. Moreover, every element $y \in \mathfrak{g}$ which commutes with x also commutes with s and n.*

One calls n the nilpotent component of x, and s its semisimple component.

Theorem 7. *Let $\phi: \mathfrak{g} \to \mathrm{End}(V)$ be a linear representation of a semisimple Lie algebra. If x is nilpotent (resp. semisimple), then so is the endomorphism $\phi(x)$.*

6. Complete Reducibility Theorem

Recall that a linear representation $\phi: \mathfrak{g} \to \operatorname{End}(V)$ is called *irreducible* (or simple) if $V \neq 0$ and if V has no invariant subspaces (submodules) other than 0 and V. One says that ϕ is *completely reducible* (or semisimple) if it is a direct sum of irreducible representations. This is equivalent to the condition that every invariant subspace of V has an invariant complement.

Theorem 8 (H. Weyl). *Every* (finite-dimensional) *linear representation of a semisimple algebra is completely reducible.*

(The algebraic proof of this theorem, to be found in Bourbaki or Jacobson, for example, is somewhat laborious. Weyl's original proof, based on the theory of compact groups (the "unitarian[1] trick") is simpler; we shall return to it later.)

7. Complex Simple Lie Algebras

The next few sections are devoted to the classification of these algebras. We will state the result straight away:

There are four series (the "four infinite families") A_n, B_n, C_n, and D_n, the index n denoting the "rank" (defined in Chapter III).

Here are their definitions:

For $n \geqslant 1$, $A_n = \mathfrak{sl}(n+1)$ is the Lie algebra of the special linear group in $n+1$ variables, $SL(n+1)$.

For $n \geqslant 2$, $B_n = \mathfrak{so}(2n+1)$ is the Lie algebra of the special orthogonal group in $2n+1$ variables, $SO(2n+1)$.

For $n \geqslant 3$, $C_n = \mathfrak{sp}(2n)$ is the Lie algebra of the symplectic group in $2n$ variables, $Sp(2n)$.

For $n \geqslant 4$, $D_n = \mathfrak{so}(2n)$ is the Lie algebra of the special orthogonal group in $2n$ variables, $SO(2n)$.

(One can also define B_n, C_n, and D_n for $n \geqslant 1$, but then:

—There are repetitions ($A_1 = B_1 = C_1$, $B_2 = C_2$, $A_3 = D_3$).
—The algebras D_1 and D_2 are not simple (D_1 is abelian and one dimensional, and D_2 is isomorphic to $A_1 \times A_1$).)

In addition to these families, there are five "exceptional" simple Lie algebras, denoted by G_2, F_4, E_6, E_7, and E_8. Their dimensions are, respectively, 14, 52, 78, 133, and 248. The algebra G_2 is the only one with a reasonably "simple" definition: it is the algebra of derivations of Cayley's octonion algebra.

---

[1] This is often referred to as the "unitary trick"; however Weyl, introducing the idea in his book "*The Classical Groups*," used the more theological word "unitarian," and we will follow him.

8. The Passage from Real to Complex

Let g_0 be a Lie algebra over $\mathbf{R}$, and $g = g_0 \otimes \mathbf{C}$ its complexification.

Theorem 9. g_0 *is abelian (resp. nilpotent, solvable, semisimple) if and only if* g *is.*

On the other hand, g_0 is simple if and only if g is simple *or* of the form $s \times \bar{s}$, with s and $\bar{s}$ simple and mutually conjugate.

Moreover, each complex simple Lie algebra g is the complexification of several nonisomorphic real simple Lie algebras; these are called the "real forms" of g. For their classification, see Séminaire S. Lie or Helgason.

CHAPTER III

Cartan Subalgebras

In this chapter (apart from Sec. 6) the ground field is the field $\mathbf{C}$ of complex numbers. The Lie algebras considered are finite dimensional.

1. Definition of Cartan Subalgebras

Let $\mathfrak{g}$ be a Lie algebra, and $\mathfrak{a}$ a subalgebra of $\mathfrak{g}$. Recall that the *normalizer* of $\mathfrak{a}$ in $\mathfrak{g}$ is defined to be the set $\mathfrak{n}(\mathfrak{a})$ of all $x \in \mathfrak{g}$ such that $\mathrm{ad}(x)(\mathfrak{a}) \subset \mathfrak{a}$; it is the largest subalgebra of $\mathfrak{g}$ which contains $\mathfrak{a}$ and in which $\mathfrak{a}$ is an ideal.

Definition 1. *A subalgebra $\mathfrak{h}$ of $\mathfrak{g}$ is called a Cartan subalgebra of $\mathfrak{g}$ if it satisfies the following two conditions:*

(a) *$\mathfrak{h}$ is nilpotent.*
(b) *$\mathfrak{h}$ is its own normalizer (that is, $\mathfrak{h} = \mathfrak{n}(\mathfrak{h})$).*

We shall see later (Sec. 3) that every Lie algebra has Cartan subalgebras.

2. Regular Elements: Rank

Let $\mathfrak{g}$ be a Lie algebra. If $x \in \mathfrak{g}$, we will let $P_x(T)$ denote the characteristic polynomial of the endomorphism $\mathrm{ad}\, x$ defined by x. We have

$$P_x(T) = \det(T - \mathrm{ad}(x)).$$

If $n = \dim \mathfrak{g}$, we can write $P_x(T)$ in the form

$$P_x(T) = \sum_{i=0}^{i=n} a_i(x)T^i.$$

If x has coordinates $x_1, \ldots, x_n$ (with respect to a fixed basis of $\mathfrak{g}$), we can view $a_i(x)$ as a function of the n complex variables $x_1, \ldots, x_n$; it is a homogeneous polynomial of degree $n - i$ in $x_1, \ldots, x_n$.

Definition 2. *The rank of $\mathfrak{g}$ is the least integer l such that the function a_l defined above is not identically zero. An element $x \in \mathfrak{g}$ is said to be regular if $a_l(x) \neq 0$.*

Remarks. Since $a_n = 1$, we must have $l \leqslant n$ with equality if and only if $\mathfrak{g}$ is nilpotent.

On the other hand, if x is a nonzero element of $\mathfrak{g}$ then $\mathrm{ad}(x)(x) = 0$, showing that 0 is an eigenvalue of $\mathrm{ad}\, x$. It follows that if $\mathfrak{g} \neq 0$ then $a_0 = 0$, so that $l \geqslant 1$.

Proposition 1. *Let $\mathfrak{g}$ be a Lie algebra. The set $\mathfrak{g}_r$ of regular elements of $\mathfrak{g}$ is a connected, dense, open subset of $\mathfrak{g}$.*

We have $\mathfrak{g}_r = \mathfrak{g} - V$, where V is defined by the vanishing of the polynomial function a_l. Clearly $\mathfrak{g}_r$ is open. Now if the interior of V were nonempty, the function a_l, vanishing on V, would be identically zero, against the definition of the rank. Finally, if $x, y \in \mathfrak{g}_r$, the (complex) line D joining x and y meets V at finitely many points. We deduce that $D \cap \mathfrak{g}_r$ is connected, and hence that x and y belong to the same connected component of $\mathfrak{g}_r$; thus $\mathfrak{g}_r$ is indeed connected.

3. The Cartan Subalgebra Associated with a Regular Element

Let x be an element of the Lie algebra $\mathfrak{g}$. If $\lambda \in \mathbf{C}$, we let $\mathfrak{g}_x^\lambda$ denote the *nilspace* of $\mathrm{ad}(x) - \lambda$; that is, the set of $y \in \mathfrak{g}$ such that $(\mathrm{ad}(x) - \lambda)^p y = 0$ for sufficiently large p.

In particular, $\mathfrak{g}_x^0$ is the nilspace of $\mathrm{ad}\, x$. Its dimension is the multiplicity of 0 as an eigenvalue of $\mathrm{ad}\, x$; that is, the least integer i such that $a_i(x) \neq 0$.

Proposition 2. *Let $x \in \mathfrak{g}$. Then:*

(a) $\mathfrak{g}$ *is the direct sum of the nilspaces $\mathfrak{g}_x^\lambda$.*
(b) $[\mathfrak{g}_x^\lambda, \mathfrak{g}_x^\mu] \subset \mathfrak{g}_x^{\lambda + \mu}$ *if $\lambda, \mu \in \mathbf{C}$.*
(c) $\mathfrak{g}_x^0$ *is a Lie subalgebra of $\mathfrak{g}$.*

Statement (a) is obtained by applying a standard property of vector space endomorphisms to $\mathrm{ad}\, x$. To prove (b), we must show that, if $y \in \mathfrak{g}_x^\lambda$ and $z \in \mathfrak{g}_x^\mu$, then $[y, z] \in \mathfrak{g}_x^{\lambda+\mu}$. Now we can use induction to prove the formula

$$(\operatorname{ad} x - \lambda - \mu)^n [y, z] = \sum_{p=0}^{n} \binom{n}{p} [(\operatorname{ad} x - \lambda)^p y, (\operatorname{ad} x - \mu)^{n-p} z].$$

If we take n sufficiently large, all terms on the right vanish, showing that $[y, z]$ is indeed in $\mathfrak{g}_x^{\lambda+\mu}$. Finally, (c) follows from (b), applied to the case $\lambda = \mu = 0$.

Theorem 1. *If x is regular, $\mathfrak{g}_x^0$ is a Cartan subalgebra of $\mathfrak{g}$; its dimension is equal to the rank l of $\mathfrak{g}$.*

First, let us show that $\mathfrak{g}_x^0$ is nilpotent. By Engel's Theorem (cf. Chapter I) it is sufficient to prove that, for each $y \in \mathfrak{g}_x^0$, the restriction of $\operatorname{ad} y$ to $\mathfrak{g}_x^0$ is nilpotent. Let $\operatorname{ad}^1 y$ denote this restriction, and $\operatorname{ad}^2 y$ the endomorphism induced by $\operatorname{ad} y$ on the quotient-space $\mathfrak{g}/\mathfrak{g}_x^0$. We put

$$U = \{y \in \mathfrak{g}_x^0 | \operatorname{ad}^1 y \text{ is not nilpotent}\}$$

$$V = \{y \in \mathfrak{g}_x^0 | \operatorname{ad}^2 y \text{ is invertible}\}.$$

The sets U and V are open in $\mathfrak{g}_x^0$. The set V is nonempty: it contains the element x. Since V is the complement of an algebraic subvariety of $\mathfrak{g}_x^0$, it follows that V is *dense* in $\mathfrak{g}_x^0$. If U were nonempty, it would therefore meet V. However, let $y \in U \cap V$. Since $y \in U$, $\operatorname{ad}^1 y$ has 0 as an eigenvalue with multiplicity strictly less than the dimension of $\mathfrak{g}_x^0$, this dimension being visibly equal to the rank l of $\mathfrak{g}$. On the other hand, since $y \in V$, 0 is not an eigenvalue of $\operatorname{ad}^2 y$. We deduce that the multiplicity of 0 as an eigenvalue of $\operatorname{ad} y$ is strictly less than l, contradicting the definition of l. Thus U is empty, and so $\mathfrak{g}_x^0$ is indeed a nilpotent algebra.

We now show that $\mathfrak{g}_x^0$ is *equal to its normalizer* $\mathfrak{n}(\mathfrak{g}_x^0)$. Let $z \in \mathfrak{n}(\mathfrak{g}_x^0)$. We have $\operatorname{ad} z(\mathfrak{g}_x^0) \subset \mathfrak{g}_x^0$, and in particular $[z, x] \in \mathfrak{g}_x^0$. By the definition of $\mathfrak{g}_x^0$, there is therefore an integer p such that $(\operatorname{ad} x)^p [z, x] = 0$, giving $(\operatorname{ad} x)^{p+1} z = 0$, so that $z \in \mathfrak{g}_x^0$ as required.

Remark. The above process provides a construction for Cartan subalgebras; we shall see that in fact it gives *all* of them.

4. Conjugacy of Cartan Subalgebras

Let $\mathfrak{g}$ be a Lie algebra. We let G denote the *inner automorphism group* of $\mathfrak{g}$; that is, the subgroup of $\operatorname{Aut}(\mathfrak{g})$ generated by the automorphisms $e^{\operatorname{ad}(y)}$ for $y \in \mathfrak{g}$.

Theorem 2. *The group G acts transitively on the set of Cartan subalgebras of $\mathfrak{g}$.*

Combining this theorem with Theorem 1, we deduce:

Corollary 1. *The dimension of a Cartan subalgebra of $\mathfrak{g}$ is equal to the rank of $\mathfrak{g}$.*

Corollary 2. *Every Cartan subalgebra of $\mathfrak{g}$ has the form $\mathfrak{g}_x^0$ for some regular element x of $\mathfrak{g}$.*

FIRST PART OF THE PROOF. In this part, $\mathfrak{h}$ denotes a Cartan subalgebra of $\mathfrak{g}$. If $x \in \mathfrak{h}$, we let $\text{ad}^1 x$ (resp. $\text{ad}^2 x$) denote the endomorphism of $\mathfrak{h}$ (resp. $\mathfrak{g}/\mathfrak{h}$) induced by x.

Lemma 1. *Let $V = \{x \in \mathfrak{h} \mid \text{ad}^2 x$ is invertible$\}$. The set V ist nonempty.*

Let us apply Lie's Theorem (cf. Chapter I) to the $\mathfrak{h}$-module $\mathfrak{g}/\mathfrak{h}$. This gives a flag:

$$0 = D_0 \subset D_1 \subset \cdots D_m = \mathfrak{g}/\mathfrak{h}$$

stable under $\mathfrak{h}$. Now $\mathfrak{h}$ acts on the one-dimensional space D_i/D_{i+1} by means of a linear form α_i:

$$\text{if } x \in \mathfrak{h}, \quad z \in D_i, \quad \text{we have } x \cdot z \equiv \alpha_i(x)z \bmod D_{i-1}.$$

(To simplify the notation, we write $x \cdot z$ instead of $\text{ad}^2 x(z)$.)

The eigenvalues of $\text{ad}^2 x$ are $\alpha_1(x), \ldots, \alpha_m(x)$. Hence it is sufficient to prove that *none of the forms α_i is identically zero*. Suppose, for example, that $\alpha_1, \ldots, \alpha_{k-1} \neq 0$ and α_k is identically zero. Let $x_0 \in \mathfrak{h}$ be chosen so that $\alpha_1(x_0) \neq 0, \ldots, \alpha_{k-1}(x_0) \neq 0$. The endomorphism of D_{k-1} (resp. of D_k) induced by $\text{ad}^2 x_0$ is invertible (resp. has 0 as an eigenvalue with multiplicity 1). The nilspace D of $\text{ad}^2 x_0$ in D_k is therefore one dimensional and is a complement for D_{k-1} in D_k. We shall show that the elements $z \in D$ are annihilated by each $\text{ad}^2 x$, $x \in \mathfrak{h}$. This is clear for x_0. Furthermore, we can use induction on n to prove the formula

$$x_0^n x \cdot z = ((\text{ad} \, x_0)^n x) \cdot z \qquad (z \in D).$$

Since the algebra $\mathfrak{h}$ is nilpotent, we have $(\text{ad} \, x_0)^n x = 0$ for sufficiently large n. This shows that $x \cdot z$ belongs to the nilspace of $\text{ad}^2 x_0$ in D_k, that is, $x \cdot z \in D$. On the other hand, $\text{ad}^2 x$ maps D_k into D_{k-1}; we therefore have $x \cdot z \in D \cap D_{k-1}$, so $x \cdot z = 0$, proving that z is indeed annihilated by each element of $\mathfrak{h}$. We now take z to be a nonzero element of D, and let $\bar{z}$ be a representative of z in $\mathfrak{g}$. The condition that $x \cdot z = 0$ for all $x \in \mathfrak{h}$ can be reinterpreted as $[x, z] \in \mathfrak{h}$ for all $x \in \mathfrak{h}$; thus z belongs to the normalizer $\mathfrak{n}(\mathfrak{h})$ of $\mathfrak{h}$. Since z is not in $\mathfrak{h}$ (because $z \neq 0$), we have $\mathfrak{n}(\mathfrak{h}) \neq \mathfrak{h}$, contradicting the definition of a Cartan subalgebra.

Lemma 2. *Let $W = G \cdot V$ be the union of the transforms of V under the action of the group G. The set W is open in $\mathfrak{g}$.*

Let $x \in V$. It is sufficient to show that W contains a neighborhood of x. Consider the map $(g, v) \mapsto g \cdot v$ from $G \times V$ to $\mathfrak{g}$, and let θ be its tangent map at the point $(1, x)$. We shall see that the image of θ is the whole of $\mathfrak{g}$. Certainly this image contains the tangent space at V, namely $\mathfrak{h}$. On the other hand, if $y \in \mathfrak{g}$ the curve

$$t \mapsto e^{t\,\mathrm{ad}(y)}x = 1 + t[y, x] + \cdots$$

has $[y, x]$ as its tangent vector at the origin. We deduce from this that $\mathrm{Im}(\mathrm{ad}\,x) \subset \mathrm{Im}(\theta)$. But since $x \in V$, $\mathrm{ad}\,x$ induces an automorphism of $\mathfrak{g}/\mathfrak{h}$, and we have

$$\mathrm{Im}(\mathrm{ad}\,x) + \mathfrak{h} = \mathfrak{g},$$

so that $\mathrm{Im}(\theta) = \mathfrak{g}$. The Implicit Function Theorem now shows that the map $G \times V \to \mathfrak{g}$ is open at the point $(1, x)$, giving the lemma.

Lemma 3. *There is a regular element x of $\mathfrak{g}$ such that $\mathfrak{h} = \mathfrak{g}_x^0$.*

Let us keep the preceding notation. Lemmas 1 and 2 show that W is open and nonempty. It therefore intersects the set $\mathfrak{g}_r$ of regular elements of $\mathfrak{g}$ (cf. Prop. 1). Now if $g \cdot x$ is regular, it is clear that x is regular. We deduce that V contains at least one regular element x. Since $\mathrm{ad}^1 x$ is nilpotent and $\mathrm{ad}^2 x$ invertible, we indeed have $\mathfrak{h} = \mathfrak{g}_x^0$.

SECOND PART OF THE PROOF. We know, thanks to Lemma 3, that the Cartan subalgebras of $\mathfrak{g}$ all have the form $\mathfrak{g}_x^0$, with $x \in \mathfrak{g}_r$. Consider the following equivalence relation R on $\mathfrak{g}_r$:

$$R(x, y) \Leftrightarrow \mathfrak{g}_x^0 \text{ and } \mathfrak{g}_y^0 \text{ are conjugate under } G.$$

Lemma 4. *The equivalence classes of R are open in $\mathfrak{g}_r$.*

We must prove that, if $x \in \mathfrak{g}_r$, every y sufficiently close to x is equivalent to x. We will apply the results of the first part of the proof to the Cartan subalgebra $\mathfrak{h} = \mathfrak{g}_x^0$. The corresponding set V contains x. By Lemma 2, $G \cdot V$ is open. Hence each element y sufficiently close to x has the form $g \cdot x'$, with $g \in G$ and $x' \in V$. We then have $\mathfrak{g}_y^0 = g \cdot \mathfrak{g}_{x'}^0 = g \cdot \mathfrak{h} = g \cdot \mathfrak{g}_x^0$, showing that x and y are indeed equivalent.

Since the equivalent classes of R are open, and since $\mathfrak{g}_r$ is connected (Prop. 1), there can be only one equivalence class. This shows that the Cartan subalgebras are indeed conjugate to each other, thus completing the proof of Theorem 2.

Remark. Theorem 2 remains true if one replaces the group G with the subgroup generated by the automorphisms of the form $e^{\mathrm{ad}(y)}$ with $\mathrm{ad}(y)$ *nilpotent*. This form of the theorem has been extended by Chevalley to the case of an arbitrary algebraically closed base field (of characteristic zero). See exposé 15 of Séminaire Sophus Lie, as well as Bourbaki, Chap. VII, Sec. 3.

5. The Semisimple Case

Theorem 3. *Let $\mathfrak{h}$ be a Cartan subalgebra of a semisimple Lie algebra $\mathfrak{g}$. Then:*

(a) *$\mathfrak{h}$ is abelian.*
(b) *The centralizer of $\mathfrak{h}$ is $\mathfrak{h}$.*
(c) *Every element of $\mathfrak{h}$ is semisimple (cf. Sec. II.5).*
(d) *The restriction of the Killing form of $\mathfrak{g}$ to $\mathfrak{h}$ is nondegenerate.*

(d) By Corollary 2 to Theorem 2, there is a regular element x such that $h = \mathfrak{g}_x^0$. Let

$$\mathfrak{g} = \mathfrak{g}_x^0 \oplus \sum_{\lambda \neq 0} \mathfrak{g}_x^\lambda$$

be the canonical decomposition of $\mathfrak{g}$ with respect to x (cf. Prop. 2). If B denotes the Killing form of $\mathfrak{g}$, then a simple calculation shows that $\mathfrak{g}_x^\lambda$ and $\mathfrak{g}_x^\mu$ are orthogonal with respect to B provided that $\lambda + \mu \neq 0$. We therefore have a decomposition of $\mathfrak{g}$ into mutually orthogonal subspaces

$$\mathfrak{g} = \mathfrak{g}_x^0 \oplus \sum_{\lambda \neq 0} (\mathfrak{g}_x^\lambda \oplus \mathfrak{g}_x^{-\lambda}).$$

Since B is nondegenerate, so is its restriction to each of these subspaces, giving (d) since $\mathfrak{h} = \mathfrak{g}_x^0$.

(a) By applying Cartan's criterion to $\mathfrak{h}$ and to the representation ad: $\mathfrak{h} \to \mathrm{End}(\mathfrak{g})$, we see that $\mathrm{Tr}(\mathrm{ad}\, x \circ \mathrm{ad}\, y) = 0$ for $x \in \mathfrak{h}$ and $y \in [\mathfrak{h}, \mathfrak{h}]$. In other words, $[\mathfrak{h}, \mathfrak{h}]$ is orthogonal to $\mathfrak{h}$ with respect to the Killing form B. Because of (d), this implies that $[\mathfrak{h}, \mathfrak{h}] = 0$.

(b) Being abelian, $\mathfrak{h}$ is contained in its own centralizer $\mathfrak{c}(\mathfrak{h})$. Moreover, $\mathfrak{c}(\mathfrak{h})$ is clearly contained in the normalizer $\mathfrak{n}(\mathfrak{h})$ of $\mathfrak{h}$. Since $\mathfrak{n}(\mathfrak{h}) = \mathfrak{h}$, we have $\mathfrak{c}(\mathfrak{h}) = \mathfrak{h}$.

(c) Let $x \in \mathfrak{h}$, and let s (resp. n) be its semisimple (resp. nilpotent) component (cf. Sec. II.5). If $y \in \mathfrak{h}$, then y commutes with x and hence also with s and n (Chapter II, Theorem 6). We therefore have $s, n \in \mathfrak{c}(\mathfrak{h}) = \mathfrak{h}$. However, since y and n commute and $\mathrm{ad}(n)$ is nilpotent, $\mathrm{ad}(y) \circ \mathrm{ad}(n)$ is also nilpotent and its trace $B(y, n)$ is zero. Thus n is orthogonal to every element of $\mathfrak{h}$. Since it belongs to $\mathfrak{h}$, n is zero by (d). Thus $x = s$, which shows that x is indeed semisimple.

Corollary 1. $\mathfrak{h}$ *is a maximal abelian subalgebra of $\mathfrak{g}$.*

This follows from (b).

Corollary 2. *Every regular element of $\mathfrak{g}$ is semisimple.*

This is because such an element is contained in a Cartan subalgebra of $\mathfrak{g}$.

Remark. One can show that every maximal abelian subalgebra of $\mathfrak{g}$ *consisting of semisimple elements* is a Cartan subalgebra of $\mathfrak{g}$. However, if $\mathfrak{g} \neq 0$ there are maximal abelian subalgebras of $\mathfrak{g}$ which contain nonzero nilpotent elements, and which are therefore not Cartan subalgebras.

6. Real Lie Algebras

Let $\mathfrak{g}_0$ be a Lie algebra over $\mathbf{R}$, and $\mathfrak{g}$ its complexification. The concepts of Cartan subalgebra, regular element, and rank are defined for $\mathfrak{g}_0$ as in the complex case. Moreover, the rank of $\mathfrak{g}_0$ is equal to that of $\mathfrak{g}$; a subalgebra $\mathfrak{h}_0$ of $\mathfrak{g}_0$ is a Cartan subalgebra if and only if its complexification $\mathfrak{h}$ is a Cartan subalgebra of $\mathfrak{g}$; an element of $\mathfrak{g}_0$ is regular in $\mathfrak{g}_0$ if and only if it is so in $\mathfrak{g}$. Theorems 1 and 3 remain true (in particular, showing the existence of Cartan subalgebras). However, this does not apply to Theorem 2: all one can say is that the Cartan subalgebras of $\mathfrak{g}_0$ are divided into finitely many classes modulo the inner automorphisms of $\mathfrak{g}_0$. (This is because the set of regular elements of $\mathfrak{g}_0$ is not necessarily connected, but rather a finite union of connected open sets.) A precise description of these classes is to be found in B. Kostant, *Proc. Nat. Acad. Sci. USA*, 1955. For more details on Cartan subalgebras, see Bourbaki, Chapter 7.

CHAPTER IV

The Algebra $\mathfrak{sl}_2$ and Its Representations

In this chapter (apart from Sec. 6) the ground field is the field $\mathbf{C}$ of complex numbers.

1. The Lie Algebra $\mathfrak{sl}_2$

This is the algebra of square matrices of order 2 and trace zero. We shall denote it by $\mathfrak{g}$. One can easily verify that it is a simple algebra, of rank 1. It has as a basis the three elements

$$X = \begin{pmatrix} 0 & 1 \\ 0 & 0 \end{pmatrix}, \quad H = \begin{pmatrix} 1 & 0 \\ 0 & -1 \end{pmatrix}, \quad Y = \begin{pmatrix} 0 & 0 \\ 1 & 0 \end{pmatrix}.$$

We have

$$[X, Y] = H, \quad [H, X] = 2X, \quad [H, Y] = -2Y.$$

The endomorphism $\mathrm{ad}(H)$ has three eigenvalues: 2, 0, -2. It follows that H is semisimple; the line $\mathfrak{h} = \mathbf{C} \cdot H$ spanned by H is a Cartan subalgebra of $\mathfrak{g}$, called the canonical Cartan subalgebra.

The elements X, Y are nilpotent. The subalgebra $\mathfrak{b}$ of $\mathfrak{g}$ generated by H and X is solvable; this is the canonical Borel subalgebra of $\mathfrak{g}$.

2. Modules, Weights, Primitive Elements

Let V be a $\mathfrak{g}$-module (not necessarily finite-dimensional). If $\lambda \in \mathbf{C}$, we will let V^λ denote the eigenspace of H in V corresponding to λ; that is, the set of all $x \in V$ such that $Hx = \lambda x$. An element of V^λ is said to have weight λ.

Proposition 1. (a) *The sum $\sum_{\lambda \in C} V^\lambda$ is direct.* (b) *If x has weight λ, then Xx has weight $\lambda + 2$ and Yx has weight $\lambda - 2$.*

(a) merely expresses the well-known fact that the eigenvectors corresponding to distinct eigenvalues are linearly independent.

Moreover, if $Hx = \lambda x$ we have

$$HXx = [H, X]x + XHx = 2Xx + \lambda Xx = (\lambda + 2)Xx,$$

and so Xx has weight $\lambda + 2$. A similar argument applies to Yx.

Remark. When V is finite dimensional, the sum $\sum V^\lambda$ is equal to V (this follows, for example, from the fact that H is semisimple; cf. Chapter II, Theorem 7). This is no longer true when V is infinite dimensional.

Definition 1. *Let V be a $\mathfrak{g}$-module, and let $\lambda \in C$. An element $e \in V$ is said to be primitive of weight λ if it is nonzero and if we have*

$$Xe = 0, \quad He = \lambda e.$$

Proposition 2. *For a nonzero element e of the $\mathfrak{g}$-module V to be primitive, it is necessary and sufficient that the line it spans should be stable under the Borel algebra $\mathfrak{b}$.*

This condition is clearly necessary. Conversely, if Ce is stable under $\mathfrak{b}$ then we have $Xe = \mu e$, $He = \lambda e$, with $\lambda, \mu \in C$. Using the formula $[H, X] = 2X$, we see that $2\mu = 0$, so $\mu = 0$ and e is indeed primitive.

Proposition 3. *Every nonzero finite-dimensional $\mathfrak{g}$-module contains a primitive element.*

This follows from Lie's Theorem (cf. Chapter I, Theorem 2′).

(Alternative proof: one chooses an eigenvector x for H, and takes the last nonzero term in the sequence $x, Xx, X^2x, \ldots$. This is a primitive element.)

3. Structure of the Submodule Generated by a Primitive Element

Theorem 1. *Let V be a $\mathfrak{g}$-module and $e \in V$ a primitive element of weight λ. Let us put $e_n = Y^n e/n!$ for $n \geqslant 0$, and $e_{-1} = 0$. Then we have*

$$\text{(i) } He_n = (\lambda - 2n)e_n$$

$$\text{(ii) } Ye_n = (n + 1)e_{n+1}$$

$$\text{(iii) } Xe_n = (\lambda - n + 1)e_{n-1}$$

for all $n \geqslant 0$.

Formula (i) asserts that e_n has weight $\lambda - 2n$, which follows from Prop. 1.
Formula (ii) is obvious.

Formula (iii) is proved by induction on n (the case $n = 0$ being true because of the convention that $e_{-1} = 0$); for we have

$$nXe_n = XYe_{n-1} = [X, Y]e_{n-1} + YXe_{n-1} = He_{n-1} + (\lambda - n + 2)Ye_{n-2}$$
$$= (\lambda - 2n + 2 + (\lambda - n + 2)(n - 1))e_{n-1}$$
$$= n(\lambda - n + 1)e_{n-1},$$

which gives (iii) on dividing by n.

Corollary 1. *Only two cases arise: either*

(a) *the elements (e_n), $n \geqslant 0$, are all linearly independent,*

or

(b) *the weight λ of e is an integer $m \geqslant 0$, the elements $e_0, \ldots, e_m$ are linearly independent, and $e_i = 0$ for $i > m$.*

Since the elements e_i have distinct weights, those which are nonzero are linearly independent (cf. Prop. 1). If they are all nonzero, then we have case (a). Otherwise, there is an integer $m \geqslant 0$ such that $e_0, \ldots, e_m$ are nonzero, and $e_{m+1} = e_{m+2} = \cdots = 0$. Applying formula (iii) with $n = m + 1$, we obtain

$$Xe_{m+1} = (\lambda - m)e_m.$$

However, $e_{m+1} = 0$ and $e_m \neq 0$. The above formula therefore implies that $\lambda = m$, so we are in case (b).

Corollary 2. *Suppose that V is finite dimensional. Then we are in case* (b) *of Corollary 1. The vector subspace W of V with basis $e_0, \ldots, e_m$ is stable under $\mathfrak{g}$; it is an irreducible $\mathfrak{g}$-module.*

Clearly case (a) of Corollary 1 is impossible. On the other hand, formulae (i), (ii), and (iii) show that W is a $\mathfrak{g}$-submodule of V (it is the $\mathfrak{g}$-submodule generated by e). By (i), the eigenvalues of H on W are equal to $m, m - 2, m - 4$, $\ldots, -m$, and have multiplicity 1. If W' is a nonzero subspace of W stable under H, then it contains one of the eigenvectors e_i ($0 \leqslant i \leqslant m$); however, if W' is stable under $\mathfrak{g}$, formulae (iii) show that W' contains $e_{i-1}, \ldots, e_0 = e$, and formulae (ii) show that it contains $e_i, e_{i+1}, \ldots$. We therefore have $W' = W$, proving the irreducibility of W.

4. The Modules W_m

Let m be an integer $\geqslant 0$, and let W_m be a vector space of dimension $m + 1$, with basis $e_0, \ldots, e_m$. Let us define endomorphisms X, Y, H of W_m by the following formulae (with the convention that $e_{-1} = e_{m+1} = 0$):

(i) $He_n = (m - 2n)e_n$

(ii) $Ye_n = (n + 1)e_{n+1}$

(iii) $Xe_n = (m - n + 1)e_{n-1}$.

A direct computation shows that

$$HXe_n - XHe_n = 2Xe_n, \quad HYe_n - YHe_n = -2Ye_n, \quad XYe_n - YXe_n = He_n,$$

in other words the endomorphisms X, Y, H make W_m into a g-module.

Theorem 2. (a) W_m *is an irreducible* g-*module.* (b) *Every irreducible* g-*module of dimension* $m + 1$ *is isomorphic to* W_m.

(a) follows from Corollary 2 to Theorem 1, and the fact that W_m is generated by the images of the primitive element e_0, which has weight m.

Let V be an irreducible g-module of dimension $m + 1$. By Prop. 3, V contains a primitive element e. Corollary 2 to Theorem 1 shows that the weight of e is an integer $m' \geqslant 0$, and that the g-submodule W of V generated by e has dimension $m' + 1$. Since V is irreducible, we must have $W = V$, so that $m' = m$, and the formulae of Theorem 1 show that V is isomorphic to W_m, as required.

EXAMPLES. The module W_0 is the trivial g-module of dimension 1. The space $\mathbf{C}^2$ with its natural g-module structure is isomorphic to W_1. The algebra g, regarded as a g-module by means of the adjoint representation, is isomorphic to W_2.

Remark. One can show that W_m is isomorphic to the *m-th symmetric power* of the module $W_1 = \mathbf{C}^2$.

5. Structure of the Finite-Dimensional g-Modules

Theorem 3. *Each finite-dimensional* g-*module is isomorphic to a direct sum of modules* W_m.

Indeed, by H. Weyl's theorem (Chapter II, Theorem 8), such a module is a direct sum of irreducible modules, and we have just seen that each finite-dimensional irreducible g-module is isomorphic to some W_m.

Theorem 4. *Let V be a finite-dimensional* g-*module. Then:*

(a) *The endomorphism of V induced by H is diagonalizable. Its eigenvalues are integers. If $\pm n$ (with $n \geqslant 0$) is an eigenvalue of H, then so are $n - 2$, $n - 4$, $\ldots, -n$.*

(b) *If n is an integer $\geqslant 0$, the linear maps*

$$Y^n: V^n \to V^{-n} \text{ and } X^n: V^{-n} \to V^n$$

are isomorphisms. In particular, V^n and V^{-n} have the same dimension. (Recall that V^n denotes the set of elements of V of weight n.)

By Theorem 3, we may assume that V is one of the $\mathfrak{g}$-modules W_m, in which case (a) and (b) are clear. ·

Remarks. (1) The fact that V^n and V^{-n} have the same dimension can also be seen by using the endomorphism $\theta = e^X e^{-Y} e^X$ of V (notice that X and Y are nilpotent on V, so that their exponentials are just polynomials). Now one checks that:

$$\theta \circ X = -Y \circ \theta, \quad \theta \circ Y = -X \circ \theta, \quad \theta \circ H = -H \circ \theta,$$

and the last identity shows that θ maps V^n to V^{-n}.

(2) Here is an example of an application of Theorems 3 and 4, independent of the interpretation of $\mathfrak{sl}_2$ as the Lie algebra of SL_2:

Let U be a compact Kähler variety of complex dimension n, and let V be the cohomology algebra $H^*(U, \mathbf{C})$. Hodge theory associates endomorphisms Λ and L of V with the kählerian structure on U (cf. A. Weil, *Variétés kähleriennes*, Chap. IV); let us take X and Y to be these endomorphisms, and define H by the relation $Hx = (n - p)x$ if $x \in H^p(U, \mathbf{C})$. Then one can check (Weil, *loc. cit.*) that V becomes a $\mathfrak{g}$-module. By applying Theorems 3 and 4 to this module, one retrieves Hodge's theorem's on "primitive" cohomology classes.

6. Topological Properties of the Group SL$_2$

This is the group of complex matrices of order 2 and determinant equal to 1. It is a complex Lie group, with Lie algebra $\mathfrak{sl}_2$. The elements X, Y, H of $\mathfrak{sl}_2$ generate the following one-parameter subgroups:

$$e^{tX} = \begin{pmatrix} 1 & t \\ 0 & 1 \end{pmatrix}, \quad e^{tY} = \begin{pmatrix} 1 & 0 \\ t & 1 \end{pmatrix}, \quad e^{tH} = \begin{pmatrix} e^t & 0 \\ 0 & e^{-t} \end{pmatrix}.$$

Similarly we will consider the subgroup SU_2 of SL_2 formed by the unitary matrices; its Lie algebra will be denoted by $\mathfrak{su}_2$.

Theorem 5. (a) SL_2 *is isomorphic* (as a real analytic variety) *to* $SU_2 \times \mathbf{R}^3$.

(b) SU_2 *is isomorphic* (as a Lie group) *to the group of quaternions of norm 1, which is itself homeomorphic to the sphere* $\mathbf{S}_3$.

(c) SU_2 *and* SL_2 *are connected and simply connected.*

(d) *The algebra* $\mathfrak{sl}_2$ *can be identified with the complexification of the real Lie algebra* $\mathfrak{su}_2$: *we have* $\mathfrak{sl}_2 = \mathfrak{su}_2 \oplus i \cdot \mathfrak{su}_2$.

The algebra $\mathfrak{su}_2$ consists of the skew hermitian matrices of order 2 and trace zero; if P denotes the set of hermitian matrices of trace zero, then clearly $P = i \cdot \mathfrak{su}_2$ and $\mathfrak{sl}_2 = \mathfrak{su}_2 \oplus P$, giving (d).

Moreover, it is straightforward to check that the map

$$(u, p) \mapsto u \cdot e^p$$

is an isomorphism (of real analytic varieties) from $SU_2 \times P$ onto SL_2. Since P is isomorphic to $\mathbf{R}^3$, this proves (a).

Statement (b) is well-known, and (c) follows from (b) and the fact that S_3 is connected and simply connected.

Now Weyl's "unitarian trick" takes the following form:

Theorem 6. *For each complex Lie group G, with Lie algebra $\mathfrak{g}$, the following canonical maps are bijections:*

$$
\begin{array}{ccc}
\mathrm{Hom}_{\mathbf{C}}(SL_2, G) & \xrightarrow{\;a\;} & \mathrm{Hom}_{\mathbf{R}}(SU_2, G) \\
\Big\downarrow{\scriptstyle b} & & \Big\downarrow{\scriptstyle c} \\
\mathrm{Hom}_{\mathbf{C}}(\mathfrak{sl}_2, \mathfrak{g}) & \xrightarrow{\;d\;} & \mathrm{Hom}_{\mathbf{R}}(\mathfrak{su}_2, \mathfrak{g}).
\end{array}
$$

(Notation: $\mathrm{Hom}_{\mathbf{C}}(SL_2, G)$ denotes the set of complex analytic homomorphisms from SL_2 to G, $\mathrm{Hom}_{\mathbf{R}}(\mathfrak{su}_2, \mathfrak{g})$ denotes the set of $\mathbf{R}$-homomorphisms from the Lie algebra $\mathfrak{su}_2$ to the Lie algebra $\mathfrak{g}$, etc. The maps a and d are the restriction maps; the maps b and c arise from the functor "Lie group" $\mapsto$ "Lie algebra".

PROOF. The maps b and c are bijective because SL_2 and SU_2 are connected and simply connected; the map d is bijective because $\mathfrak{sl}_2$ is the complexification of $\mathfrak{su}_2$; the bijectivity of a (which is not *a priori* obvious) follows from the commutativity of the diagram.   □

Corollary. *The finite-dimensional linear representations of* SU_2, SL_2, $\mathfrak{su}_2$, *and* $\mathfrak{sl}_2$ *correspond bijectively with each other.*

It is sufficient to apply the theorem to the group $G = GL_n(\mathbf{C})$ for $n = 0$, $1, \ldots$.

7. Applications

The global results of the preceding section provide alternative proofs of certain properties of $\mathfrak{sl}_2$-modules. Thus, for example:

(i) The *complete reducibility* of finite-dimensional $\mathfrak{sl}_2$-modules follows from the fact that, by the corollary to Theorem 6, these modules correspond to the linear representations of the *compact* group SU_2.

(ii) The fact that the eigenvalues of H are integers can be seen in the following way: let V be a finite-dimensional $\mathfrak{sl}_2$-module, and let $x \in V$ be an eigenvector of H, with eigenvalue λ. By the corollary to Theorem 6, the group SL_2 acts on V; in particular, the element e^{tH} of SL_2 sends x to $e^{t\lambda x}$; but, if $t = 2i\pi$, we have $e^{tH} = 1$ in SL_2, so $e^{tH}x = x$. We must therefore have $e^{t\lambda} = 1$ for $t = 2i\pi$, implying that λ is an *integer*.

(iii) The automorphism θ introduced at the end of Sec. 5 corresponds to the action of the element $\begin{pmatrix} 0 & 1 \\ -1 & 0 \end{pmatrix}$ of SL_2.

CHAPTER V

Root Systems

In this chapter (apart from Sec. 17) the ground field is the field $\mathbf{R}$ of real numbers. The vector spaces considered are all finite dimensional.

1. Symmetries

Let V be a vector space and α a nonzero element of V. One defines a *symmetry with vector* α to be any automorphism s of V satisfying the following two conditions:

(i) $s(\alpha) = -\alpha$.
(ii) The set H of elements of V fixed by s is a hyperplane of V.

It is clear that H is then a complement for the line $\mathbf{R}\alpha$ spanned by α, and that s has order 2. The symmetry s is completely determined by the choice of $\mathbf{R}\alpha$ and of H.

Let V^* be the dual space of V, and let α^* be the unique element of V^* which vanishes on H and takes the value 2 on α. We have

$$s(x) = x - \langle \alpha^*, x \rangle \alpha \qquad \text{for all } x \in V,$$

which we can write as

$$s = 1 - \alpha^* \otimes \alpha,$$

on identifying $\text{End}(V)$ and $V^* \otimes V$.

Conversely, if $\alpha \in V$ and $\alpha^* \in V^*$ satisfy

$$\langle \alpha^*, \alpha \rangle = 2,$$

the element $1 - \alpha^* \otimes \alpha$ is a symmetry with vector α.

Lemma. *Let α be a nonzero element of V, and let R be a finite subset of V which spans V. There is at most one symmetry with vector α which leaves R invariant.*

Let s and s' be two such symmetries, and let u be their product. The automorphism u has the following properties:

$$u(R) = R,$$

$$u(\alpha) = \alpha,$$

u induces the identity on $V/\mathbf{R}\alpha$.

The last two properties show that the eigenvalues of u are equal to 1. Moreover, because R is finite there is an integer $n \geqslant 1$ such that $u^n(x) = x$ for all $x \in R$, so that $u^n = 1$ since R spans V. This implies that u is diagonalizable. Since its eigenvalues are equal to 1, we therefore have $u = 1$, so that $s = s'$.

2. Definition of Root Systems

Definition 1. *A subset R of a vector space V is said to be a root system in V if the following conditions are satisfied:*

(1) *R is finite, spans V, and does not contain 0.*
(2) *For each $\alpha \in R$, there is a symmetry s_α, with vector α, leaving R invariant.*
   (This symmetry is unique, by Lemma 1.)
(3) *For each $\alpha, \beta \in R$, $s_\alpha(\beta) - \beta$ is an integer multiple of α.*

The dimension of V is called the *rank* of R. The elements of R are called the *roots* of V (relative to R). By Sec. 1, the symmetry s_α associated with the root α can be written uniquely as

$$s_\alpha = 1 - \alpha^* \otimes \alpha \qquad \text{with } \langle \alpha^*, \alpha \rangle = 2.$$

The element α^* of V^* is called the *inverse root* of α. Condition (3) is equivalent to the following:

(3') *For all $\alpha, \beta \in R$, we have $\langle \alpha^*, \beta \rangle \in \mathbf{Z}$.*

Let $\alpha \in R$. By (2) and (3), we have $-\alpha \in R$, since $-\alpha = s_\alpha(\alpha)$.

Definition 2. *A root system R is said to be reduced if, for each $\alpha \in R$, α and $-\alpha$ are the only roots proportional to α.*

If a root system R is not reduced, it contains two proportional roots α and $t\alpha$, with $0 < t < 1$. Applying (3) to $\beta = t\alpha$, we see that $2t \in \mathbf{Z}$, which implies that $t = \frac{1}{2}$.

Then the roots proportional to α are simply

$$-\alpha, \quad -\alpha/2, \quad \alpha/2, \quad \alpha.$$

Remark. The reduced roots systems are those which arise in the theory of semisimple Lie algebras (or algebraic groups) over an algebraically closed field; they are the only ones we shall need. Nonreduced systems occur when one no longer assumes that the base field is algebraically closed.

3. First Examples

(We shall see others in Sec. 16.) Clearly, the only reduced system of rank 1 is the system

   (denoted by A_1).

There is one nonreduced system of rank 1:

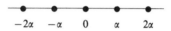

One can show (cf. Secs. 8, 15) that every reduced system of rank 2 is isomorphic to one of the following four:

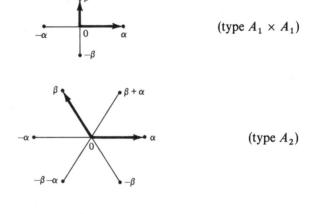

(type $A_1 \times A_1$)

(type A_2)

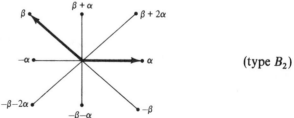

(type B_2)

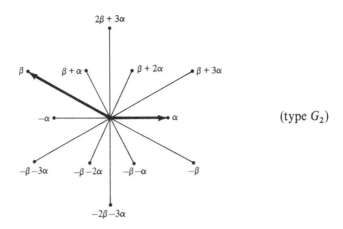

$$2\beta + 3\alpha$$

$$\beta \qquad \beta + \alpha \qquad \beta + 2\alpha \qquad \beta + 3\alpha$$

$$-\alpha \qquad \alpha \qquad \text{(type } G_2\text{)}$$

$$-\beta - 3\alpha \qquad -\beta - 2\alpha \qquad -\beta - \alpha \qquad -\beta$$

$$-2\beta - 3\alpha$$

EXERCISE. Complete the root system B_2 so as to obtain a nonreduced system. Can one do the same with A_2 and G_2?

4. The Weyl Group

Definition 3. *Let R be a root system in a vector space V. The Weyl group of R is the subgroup W of $\mathrm{GL}(V)$ generated by the symmetries s_α, $\alpha \in R$.*

The group W is a normal subgroup of the group $\mathrm{Aut}(R)$ of automorphisms of V leaving R invariant. Since R spans V, these two groups can be identified with subgroups of the group of all permutations of R; they are *finite* groups.

EXAMPLE. When R is a reduced system of rank 2, the group W is isomorphic to the dihedral group of order $2n$, with $n = 2$ (type $A_1 \times A_1$), $n = 3$ (type A_2), $n = 4$ (type B_2), or $n = 6$ (type G_2). We have $\mathrm{Aut}(R) = W$ when R is of type B_2 or G_2, and $|\mathrm{Aut}(R): W| = 2$ when R is of type $A_1 \times A_1$ or A_2.

5. Invariant Quadratic Forms

Proposition 1. *Let R be a root system in V. There is a positive definite symmetric bilinear form $(,)$ on V which is invariant under the Weyl group W of R.*

This follows simply from the fact that W is finite. For if $B(x, y)$ is any positive definite symmetric bilinear form on V, the form

$$(x, y) = \sum_{w \in W} B(wx, wy)$$

is invariant, and $(x, x) > 0$ for all $x \neq 0$.

From now onwards, we let $(,)$ denote such a form. The choice of $(,)$ gives V the structure of a *Euclidean space*, with respect to which the elements of W are *orthogonal* transformations. In particular this applies to the symmetries s_α; we deduce from this that we have

$$s_\alpha(x) = x - 2\frac{(x, \alpha)}{(\alpha, \alpha)}\alpha \qquad \text{for all } x \in V.$$

Let α' be the element of V corresponding to α^* under the isomorphism $V \to V^*$ determined by the chosen bilinear form. By definition, we have

$$s_\alpha(x) = x - (\alpha', x)\alpha \qquad \text{for all } x \in V.$$

Comparing this with the preceding formula, we get

$$\alpha' = \frac{2\alpha}{(\alpha, \alpha)}.$$

(Thus we pass from α to α' by an "inversion in a sphere of radius $\sqrt{2}$," in the sense of elementary geometry.)

Condition (3) for root systems can be written as

$$2\frac{(\alpha, \beta)}{(\alpha, \alpha)} \in \mathbf{Z} \qquad \text{for } \alpha, \beta \in R.$$

Thus one can retrieve the traditional definition of root systems, cf. Jacobson or Séminaire S. Lie. (The definition in Sec. 2 is that of Bourbaki, *Systèmes de Racines*—it has the advantage of separating the roles of V and of V^*.)

6. Inverse Systems

Let R be a root system in V.

Proposition 2. *The set R^* of inverse roots α^*, $\alpha \in R$, is a root system in V^*. Moreover, $\alpha^{**} = \alpha$ for all $\alpha \in R$.*

Clearly R^* is finite and does not contain 0. To prove that it spans V^* it is sufficient (by the isomorphism $V \to V^*$) to show that the elements $\alpha' = 2\alpha/(\alpha, \alpha)$ span V, which is obvious. If $\alpha^* \in R^*$ we take the corresponding symmetry to be the transpose ${}^t s_\alpha = 1 - \alpha \otimes \alpha^*$ of s_α. Since $s_\alpha(R) = R$, we have $s_{\alpha^*}(R^*) = R^*$. Similarly, we see that $\alpha^{**} = \alpha$. Finally, if $\alpha^*, \beta^* \in R^*$, we have

$$\langle \alpha^{**}, \beta^* \rangle = \langle \beta^*, \alpha \rangle \in \mathbf{Z},$$

as required.

The system R^* is called the *inverse* (or *dual*) *system* of the system R. Its Weyl group can be identified with that of R by means of the map

$$w \mapsto {}^t w^{-1}.$$

7. Relative Position of Two Roots

Let us keep the notation of the preceding sections. If α, β are two roots, we put

$$n(\beta, \alpha) = \langle \alpha^*, \beta \rangle = 2\frac{(\alpha, \beta)}{(\alpha, \alpha)}.$$

We have $n(\beta, \alpha) \in \mathbf{Z}$. Now if we let $|\alpha|$ denote the length of α (that is, $(\alpha, \alpha)^{1/2}$), and ϕ the angle between α and β (with respect to the Euclidean structure on V), then we have $(\alpha, \beta) = |\alpha||\beta|\cos\phi$, so that

$$n(\beta, \alpha) = 2\frac{|\beta|}{|\alpha|}\cos\phi.$$

From this we deduce the formula

$$n(\beta, \alpha)n(\alpha, \beta) = 4\cos^2\phi.$$

Since $n(\beta, \alpha)$ is an integer, $4\cos^2\phi$ can take only the values 0, 1, 2, 3, 4; the last case being that in which α and β are proportional.

Returning to the case of *nonproportional* roots, we see that there are 7 possibilities (up to transposition of α and β):

1   $n(\alpha, \beta) = 0,$ $n(\beta, \alpha) = 0,$ $\phi = \pi/2.$

2   $n(\alpha, \beta) = 1,$ $n(\beta, \alpha) = 1,$ $\phi = \pi/3,$ $|\beta| = |\alpha|.$

3   $n(\alpha, \beta) = -1,$ $n(\beta, \alpha) = -1,$ $\phi = 2\pi/3,$ $|\beta| = |\alpha|.$

4   $n(\alpha, \beta) = 1,$ $n(\beta, \alpha) = 2,$ $\phi = \pi/4,$ $|\beta| = \sqrt{2}|\alpha|.$

5   $n(\alpha, \beta) = -1,$ $n(\beta, \alpha) = -2,$ $\phi = 3\pi/4,$ $|\beta| = \sqrt{2}|\alpha|.$

6   $n(\alpha, \beta) = 1,$ $n(\beta, \alpha) = 3,$ $\phi = \pi/6,$ $|\beta| = \sqrt{3}|\alpha|.$

7   $n(\alpha, \beta) = -1,$ $n(\beta, \alpha) = -3,$ $\phi = 5\pi/6,$ $|\beta| = \sqrt{3}|\alpha|.$

Notice that knowledge of the angle ϕ determines the *set* $\{n(\alpha, \beta), n(\beta, \alpha)\}$, or, what amounts to the same thing, the set of ratios of lengths

$$\left\{\frac{|\alpha|}{|\beta|}, \frac{|\beta|}{|\alpha|}\right\},$$

provided that we have $\phi \neq \pi/2$.

Proposition 3. Let α and β be two nonproportional roots. If $n(\beta, \alpha) > 0$, then $\alpha - \beta$ is a root.

(Notice that $n(\beta, \alpha) > 0$ is equivalent to $(\alpha, \beta) > 0$: the two roots form an *acute* angle.)

The above list shows that we have either $n(\beta, \alpha) = 1$ or $n(\alpha, \beta) = 1$. In the first case,

$$\alpha - \beta = -(\beta - n(\beta, \alpha)\alpha) = -s_\alpha(\beta),$$

so that $\alpha - \beta \in R$. In the second case, $\alpha - \beta = s_\beta(\alpha) \in R$.

8. Bases

Let R be a root system in V.

Definition 4. *A subset S of R is called a base for R if the following two conditions are satisfied*:

(i) S *is a basis for the vector space V.*
(ii) *Each $\beta \in R$ can be written as a linear combination*

$$\beta = \sum_{\alpha \in S} m_\alpha \alpha,$$

where the coefficients m_α are integers with the same sign (that is, all $\geqslant 0$ or all $\leqslant 0$).

Instead of "base," the terms "simple root system" or "fundamental root system" are also used; the elements of S are then called the "simple roots."

Theorem 1. *There exists a base.*

We shall prove a more precise result.

Let $t \in V^*$ be an element such that $\langle t, \alpha \rangle \neq 0$ for all $\alpha \in R$. Let R_t^+ be the set of all $\alpha \in R$ such that $\langle t, \alpha \rangle$ is > 0; we have $R = R_t^+ \cup (-R_t^+)$. An element α of R_t^+ is called *decomposable* if there exist $\beta, \gamma \in R_t^+$ such that $\alpha = \beta + \gamma$; otherwise, α is called *indecomposable*. Let S_t be the set of indecomposable elements of R_t^+.

Proposition 4. *S_t is a base for R. Conversely, if S is a base for R, and if $t \in V^*$ is such that $\langle t, \alpha \rangle$ is > 0 for all $\alpha \in S$, then $S = S_t$.*

Let us show that S_t is a base for R. We will do this in stages.

Lemma 2. *Each element of R_t^+ is a linear combination, with non-negative integer coefficients, of elements of S_t.*

Let I be the set of $\alpha \in R_t^+$ which do not have the property in question. If I were nonempty, there would be an element $\alpha \in I$ with $\langle t, \alpha \rangle$ minimal. The element α is decomposable (otherwise it would belong to S_t); if we write $\alpha = \beta + \gamma$, with $\beta, \gamma \in R_t^+$, we have

$$\langle t, \alpha \rangle = \langle t, \beta \rangle + \langle t, \gamma \rangle,$$

and since $\langle t, \beta \rangle$ and $\langle t, \gamma \rangle$ are > 0, they are strictly smaller than $\langle t, \alpha \rangle$. We therefore have $\beta \notin I$ and $\gamma \notin I$, clearly giving $\alpha \notin I$, a contradiction.

Lemma 3. *We have $(\alpha, \beta) \leqslant 0$ if $\alpha, \beta \in S_t$.*

Otherwise, Prop. 3 would show that $\gamma = \alpha - \beta$ is a root. We would then have either $\gamma \in R_t^+$, in which case $\alpha = \beta + \gamma$ would be decomposable, or else $-\gamma \in R_t^+$, in which case $\beta = \alpha + (-\gamma)$ would be decomposable.

Lemma 4. *Let* $t \in V^*$ *and* $A \subset V$ *be such that:*

$$\text{(a)} \quad \langle t, \alpha \rangle > 0 \qquad \text{for all } \alpha \in A,$$

$$\text{(b)} \quad (\alpha, \beta) \leqslant 0 \qquad \text{for all } \alpha, \beta \in A.$$

Then the elements of A *are linearly independent.*

(In other words, vectors which subtend obtuse angles with each other, and lie in the same half-space, are linearly independent.)

Each relation between the elements of A can be written in the form

$$\sum y_\beta \beta = \sum z_\gamma \gamma,$$

where the coefficients y_β and z_γ are $\geqslant 0$, and where β and γ range over *disjoint* finite subsets of A.

Let $\lambda \in V$ be the element $\sum y_\beta \beta$. We have

$$(\lambda, \lambda) = \sum y_\beta z_\gamma (\beta, \gamma),$$

so that $(\lambda, \lambda) \leqslant 0$, by (b).

We deduce that $\lambda = 0$. But then we have

$$0 = \langle t, \lambda \rangle = \sum y_\beta \langle t, \beta \rangle,$$

so that $y_\beta = 0$ for all β, and similarly $z_\gamma = 0$ for all γ, as required.

Lemmas 2, 3, 4 show that S_t is a base for R. Conversely, let S be a base for R, and let $t \in V^*$ be such that $\langle t, \alpha \rangle$ is >0 for all $\alpha \in S$. If we let R^+ denote the set of linear combinations with non-negative integer coefficients of elements of S, then we have $R^+ \subset R_t^+$ and $(-R^+) \subset (-R_t^+)$, so that $R^+ = R_t^+$ since R is the union of R^+ and $-R^+$. We deduce that the elements of S are indecomposable in R_t^+, that is, that $S \subset S_t$. Since S and S_t have the same number of elements (namely the dimension of V), we have $S = S_t$.

EXAMPLE. Suppose that dim $V = 2$, and let $\{\alpha, \beta\}$ be a base for R. Since the angle between α and β is obtuse (Lemma 3), only cases 1/, 3/, 5/, 7/ of Sec. 7 are possible (allowing for a possible transposition of α and β). They correspond to the systems of types $A_1 \times A_1$, A_2, B_2, and G_2 respectively (cf. Sec. 3).

9. Some Properties of Bases

In the following sections, S denotes a *base* for the root system R. We denote by R^+ the set of roots which are linear combinations with non-negative integer coefficients, of elements of S. An element of R^+ is called a *positive root* (with respect to S).

Proposition 5. *Every positive root β can be written as*

$$\beta = \alpha_1 + \cdots + \alpha_k \qquad with \ \alpha_i \in S,$$

in such a way that the partial sums

$$\alpha_1 + \cdots + \alpha_h, \qquad 1 \leqslant h \leqslant k,$$

are all roots.

Let $t \in V^*$ satisfy $\langle t, \alpha \rangle = 1$ for all $\alpha \in S$. Since β is a positive root, $\langle t, \beta \rangle$ is a non-negative integer. We shall prove the proposition by induction on $k = \langle t, \beta \rangle$. First note that the values of (α, β), $\beta \in S$, cannot all be $\leqslant 0$. If they were, then Lemma 4 would show that β and the elements of S were linearly independent, which is absurd. Hence there exists some $\alpha \in S$ such that $(\alpha, \beta) > 0$. If α and β are proportional, we have $\beta = \alpha$ or $\beta = 2\alpha$, and Prop. 5 is true. Otherwise, Prop. 3 shows that $\gamma = \beta - \alpha$ is a root. If $\gamma \in -R^+$, $\alpha = \beta + (-\gamma)$ would be decomposable, which is absurd. Hence we have $\gamma \in R^+$ and $\langle t, \gamma \rangle = k - 1$. The induction hypothesis can be applied to γ, so we obtain the result by taking $\alpha_k = \alpha$.

Proposition 6. *Suppose that R is reduced, and let $\alpha \in S$. The symmetry s_α associated with α leaves $R^+ - \{\alpha\}$ invariant.*

Let $\beta \in R^+ - \{\alpha\}$. We have

$$\beta = \sum_{\gamma \in S} m_\gamma \gamma \qquad with \ m_\gamma \geqslant 0.$$

Since R is reduced, and $\beta \neq \alpha$, β is not proportional to α, and there exists some $\gamma \neq \alpha$ such that $m_\gamma \neq 0$. Since $s_\alpha(\beta) = \beta - n(\beta, \alpha)\alpha$, we see that the coefficient of γ in $s_\alpha(\beta)$ is equal to m_γ. This gives $s_\alpha(\beta) \in R^+$, proving the proposition.

Corollary. *Let ρ be half the sum of the positive roots. We have*

$$s_\alpha(\rho) = \rho - \alpha$$

for all $\alpha \in s$.

Let ρ_α be half the sum of the elements of $R^+ - \{\alpha\}$. Clearly we have $s_\alpha(\rho_\alpha) = \rho_\alpha$. On the other hand, $\rho = \rho_\alpha + \alpha/2$. Since $s_\alpha(\alpha) = -\alpha$ we deduce from this that $s_\alpha(\rho) = \rho - \alpha$.

Proposition 7. *Suppose that R is reduced. The set S^* of inverse roots of the elements of S is a base for R^*.*

Let R' be the root system consisting of the vectors $\alpha' = 2\alpha/(\alpha, \alpha)$ for $\alpha \in R$. By the isomorphism $V \to V^*$ (cf. Sec. 5) it is sufficient to prove that the vectors α', $\alpha \in S$, form a base for R'. If $t \in V^*$ is such that $\langle t, \alpha \rangle > 0$ for all $\alpha \in S$, then

$(R')_t^+$ consists of the vectors α' with $\alpha \in R^+$. The convex cone C generated by $(R')_t^+$ is therefore the same as that generated by R^+. Let S'_t be the corresponding base of R'. The half-lines generated by the elements of S'_t are the *extremal generators* of C; hence they are the half-lines $\mathbf{R}^+\alpha$, with $\alpha \in S$. Since R is reduced, such a half-line contains a unique root of R', which must be α'. Thus $S'_t = S$, as required.

Remark. In the general case, let S_1 (resp. S_2) be the subset of S consisting of the roots α such that 2α is not a root (resp. 2α is a root). We obtain a base for R^* by taking the elements α^*, $\alpha \in S_1$, and the elements $\alpha^*/2$, $\alpha \in S_2$.

10. Relations with the Weyl Group

We assume that R is *reduced*.

Theorem 2. *Let W be the Weyl group of R.*

(a) *For each $t \in V^*$, there exists $w \in W$ such that $\langle w(t), \alpha \rangle \geq 0$ for all $\alpha \in S$.*
(b) *If S' is a base for R, there exists $w \in W$ such that $w(S') = S$.*
(c) *For each $\beta \in R$, there exists $w \in W$ such that $w(\beta) \in S$.*
(d) *The group W is generated by the symmetries s_α, $\alpha \in S$.*

Let W_S be the subgroup of W generated by the symmetries s_α, $\alpha \in S$. We first *prove* (a) *for the group W_S.* Let $t \in V^*$, and let ρ be half the sum of the positive roots (cf. Prop. 6, Corollary). Let us choose an element w of W_S so that

$$\langle w(t), \rho \rangle$$

is *maximal*. In particular, we have

$$\langle w(t), \rho \rangle \geq \langle s_\alpha w(t), \rho \rangle \qquad \text{if } \alpha \in S.$$

But we have

$$\langle s_\alpha w(t), \rho \rangle = \langle w(t), s_\alpha(\rho) \rangle = \langle w(t), \rho - \alpha \rangle$$

(cf. the corollary to Prop. 6). Hence we conclude that $\langle w(t), \alpha \rangle \geq 0$, which proves our assertion.

We now prove (b) *for the group W_S.* Let t' be an element of V^* such that $\langle t', \alpha' \rangle > 0$ for all $\alpha' \in S'$. By (a), there exists $w \in W_S$ such that, if we put $t = w(t')$, then $\langle t, \alpha \rangle \geq 0$ for all $\alpha \in S$. Since $\langle t, \alpha \rangle = \langle t', w^{-1}(\alpha) \rangle$, and since t' is not orthogonal to any root, we in fact have $\langle t, \alpha \rangle > 0$ for all $\alpha \in S$. By Prop. 4, we have

$$S = S_t \quad \text{and} \quad S' = S_{t'}.$$

Since w sends t' to t, it also sends S' to S.

We now prove (c) *for the group* W_S. Let $\beta \in R$, and let L be the hyperplane of V^* orthogonal to β. The hyperplanes associated with the roots other than $\pm\beta$ are *distinct* from L, and there are only finitely many of them. Hence there is an element t_0 of L not contained in any of these hyperplanes. We have

$$\langle t_0, \beta \rangle = 0 \quad \text{and} \quad \langle t_0, \gamma \rangle \neq 0 \qquad \text{for } \gamma \in R, \gamma \neq \pm\beta.$$

One can find an element t sufficiently close to t_0 that $\langle t, \beta \rangle = \varepsilon$, with $\varepsilon > 0$, the absolute value of each $\langle t, \gamma \rangle$, $\gamma \neq \pm\beta$, being strictly greater than ε. Let S_t be the base of R associated with t as described in Sec. 8; clearly β belongs to S_t. By (b), there exists $w \in W$ such that $w(S_t) = S$. We then have $w(\beta) \in S$.

We finally show that $W_S = W$, which will prove (d). Since W is generated by the symmetries s_β, with $\beta \in R$, it is sufficient to show that $s_\beta \in W_S$. By (c), there exists $w \in W_S$ such that $\alpha = w(\beta)$ belongs to S. We have

$$s_\alpha = s_{w(\beta)} = w \cdot s_\beta \cdot w^{-1},$$

so that $s_\beta = w^{-1} \cdot s_\alpha \cdot w$, which indeed shows that $s_\beta \in W_S$.

Remarks. (1) The element w given in (b) is unique (cf. Sec. VII.5).

(2) The set of elements $t \in V^*$ such that $\langle t, \alpha \rangle > 0$ for all $\alpha \in S$ is called the *Weyl chamber* associated with S. By (a) and (b), the Weyl chambers are the *connected components* of the complement in V^* of the hyperplanes orthogonal to the roots; the group W permutes them transitively.

(3) One can refine (d) by showing that the *relations* between the generators $s_\alpha (\alpha \in S)$ of W are all consequences of the following:

$$s_\alpha^2 = 1, \quad (s_\alpha s_\beta)^{m(\alpha, \beta)} = 1,$$

where $m(\alpha, \beta)$ is equal to 2, 3, 4, or 6 as the angle between α and β is $\pi/2$, $2\pi/3$, $3\pi/4$, or $5\pi/6$. See, for example, *Séminaire Chevalley*, 1956–58, exposé 14, or Bourbaki, Chaps. IV–V.

11. The Cartan Matrix

Definition 5. *The Cartan matrix of* R (with respect to the chosen base S) *is the matrix* $(n(\alpha, \beta))_{\alpha, \beta \in S}$.

We recall (cf. Sec. 7) that $n(\alpha, \beta) = \langle \beta^*, \alpha \rangle$ is an integer. We have $n(\alpha, \alpha) = 2$; if $\alpha \neq \beta$, we know (cf. Lemma 3) that $n(\alpha, \beta) \leq 0$. We have $n(\alpha, \beta) = 0, -1, -2$, or -3.

EXAMPLE. The Cartan matrix of G_2 is $\begin{pmatrix} 2 & -1 \\ -3 & 2 \end{pmatrix}$.

Proposition 8. *A reduced root system is determined, up to isomorphism, by its Cartan matrix.*

More precisely:

Proposition 8'. *Let R' be a reduced root system in a vector space V', let S' be a base for R', and let $\phi: S \to S'$ be a bijection such that $n(\phi(\alpha), \phi(\beta)) = n(\alpha, \beta)$ for all $\alpha, \beta \in S$. If R is reduced, then there is a unique isomorphism $f: V \to V'$ which is an extension of ϕ and maps R onto R'.*

To define f, we extend ϕ by linearity from S to V. If $\alpha, \beta \in S$, we have

$$s_{\phi(\alpha)} \circ f(\beta) = s_{\phi(\alpha)}(\phi(\beta)) = \phi(\beta) - n(\phi(\alpha), \phi(\beta))\phi(\alpha)$$

and

$$f \circ s_\alpha(\beta) = f(\beta - n(\beta, \alpha)\alpha) = \phi(\beta) - n(\beta, \alpha)\phi(\alpha).$$

Comparing these, we see that $s_{\phi(\alpha)} \circ f = f \circ s_\alpha$ for all $\alpha \in S$. If W (resp. W') denotes the Weyl group of R (resp. R'), we see that $W' = fWf^{-1}$. Since $R = W(S)$ and $R' = W(S')$, we deduce that $f(R) = R'$, as required.

In particular, let E be the group of permutations of S which leave the Cartan matrix invariant. By the above argument, E *can be identified with the group of automorphisms of R which leave the base S invariant.*

Proposition 9. *The group* $\mathrm{Aut}(R)$ *is the semidirect product of E and W.*

If $w \in W \cap E$, we have $w(S) = S$, so that $w = 1$ by a result which will be proved later (Sec. VII.5). Moreover, if $u \in \mathrm{Aut}(R)$, $u(S)$ is a base for R, hence there exists $w \in W$ such that $w(u(S)) = S$ (cf. Theorem 2). We therefore have $wu \in E$, showing that $\mathrm{Aut}(R) = W \cdot E$.

Corollary. *The group* $\mathrm{Aut}(R)/W$ *is isomorphic to E.*

12. The Coxeter Graph

Definition 6. *A Coxeter graph is a finite graph, each pair of distinct vertices being joined by 0, 1, 2, or 3 edges.*

Let R be a root system, and let S be a base for R. The Coxeter graph of R (with respect to S) is defined as follows: the vertices are the elements of S, two distinct vertices α and β being joined by 0, 1, 2, or 3 edges as $n(\alpha, \beta) \cdot n(\beta, \alpha)$ is equal to 0, 1, 2, or 3.

(Recall that if ϕ denotes the angle between α and β, then

$$n(\alpha, \beta) \cdot n(\beta, \alpha) = 4\cos^2 \phi,$$

cf. Sec. 7.)

Of course, the transitivity theorem in Sec. 10 shows that the graphs associated with different bases of R are isomorphic.

EXAMPLES. The Coxeter graphs of the root systems in Sec. 3 are the following:

$$\circ \qquad \text{type } A_1$$

$$\circ \quad \circ \quad \text{type } A_1 \times A_1$$

$$\circ\!\!-\!\!-\!\!\circ \quad \text{type } A_2$$

$$\circ\!\!=\!\!=\!\!\circ \quad \text{type } B_2$$

$$\circ\!\!\equiv\!\!\equiv\!\!\circ \quad \text{type } G_2.$$

13. Irreducible Root Systems

Proposition 10. *Suppose that V is the direct sum of two subspaces V_1 and V_2, and that R is contained in $V_1 \cup V_2$. Let $R_i = R \cap V_i$, $i = 1, 2$. Then:*

(a) *V_1 and V_2 are orthogonal.*
(b) *R_i is a root system in V_i.*

If $\alpha \in R_1$ and $\beta \in R_2$, $\alpha - \beta$ is not contained in $V_1 \cup V_2$, so it is not a root. By Prop. 3, we therefore have $(\alpha, \beta) \leqslant 0$. Since this also applies to α and $-\beta$, we see that $(\alpha, \beta) = 0$. Since R_i spans V_i, (a) follows.

For (b), it is sufficient to notice that, by (a), the symmetry associated with an element of R_1 preserves V_2, and hence also V_1.

One says that the system R is the *sum* of the subsystems R_i. If this can happen only trivially (that is, with V_1 or V_2 equal to 0), and if $V \neq 0$, then R is said to be *irreducible*.

Proposition 11. *Every root system is a sum of irreducible systems.*

This is obvious.

One can show that such a decomposition is *unique*.

Proposition 12. *For R to be irreducible, it is necessary and sufficient that its Coxeter graph should be connected and nonempty.*

If R is the sum of two nontrivial subsystems R_1 and R_2, we can take the union of two bases S_1 and S_2 for R_1 and R_2 to be a base S for R. If $\alpha \in S_1$ and $\beta \in S_2$, then α and β are orthogonal and are therefore *not joined by any edge* in the Coxeter graph of S. We deduce that the latter is the *disjoint sum* of the Coxeter graphs of the bases S_i, and is therefore not connected.

Conversely, if S has a nontrivial partition

$$S = S_1 \cup S_2$$

such that every element of S_1 is orthogonal to every element of S_2, then the vector subspaces V_1 and V_2 spanned by S_1 and S_2 are orthogonal, and are therefore invariant under the symmetries s_α, $\alpha \in S$. Hence R is contained in $V_1 \cup V_2$, and is therefore reducible.

14. Classification of Connected Coxeter Graphs

Theorem* 3. *Every connected nonempty Coxeter graph which is attached to a root system is isomorphic to one of the following:*

A_n: ○——○— $\cdots$ —○——○   (*n vertices*, $n \geqslant 1$)

B_n: ○——○— $\cdots$ —○══○   (*n vertices*, $n \geqslant 2$)

D_n: ○——○ $\cdots$ —○⟨   (*n vertices*, $n \geqslant 4$)

G_2: ○≡══○

F_4: ○——○══○——○

E_6: ○——○——○——○——○ with branch

E_7: ○——○——○——○——○——○ with branch

E_8: ○——○——○——○——○——○——○ with branch

The principle of the proof is as follows.

One takes a nonempty connected Coxeter graph G, with vertex-set S. One associates with G a symmetric bilinear form $(,)$ on the space $\mathbf{R}^S$ with basis $(e_\alpha)_{\alpha \in S}$, by defining

$$(e_\alpha, e_\alpha) = 1$$

$(e_\alpha, e_\beta) = \cos(\pi/2), \cos(2\pi/3), \cos(3\pi/4), \cos(5\pi/6)$ as α and β are joined by 0, 1, 2, or 3 edges.

For G to be the Coxeter graph of a root system, it is necessary that this form should be *positive definite* (for it can be realized by one of the invariant forms of Sec. 5). One then shows, by a series of ingenious reductions, that this positivity condition is sufficient to force an isomorphism between G and A_n, B_n, ... or E_8. For further details, see Séminaire S. Lie, exposé 13; Jacobson, pp. 128–134; or Bourbaki, Chap. 6, Sec. 4.

15. Dynkin Diagrams

(For simplicity, we restrict our attention to root systems which are both reduced and irreducible.)

The Coxeter graph *is not sufficient* to determine the Cartan matrix (and hence the root system); indeed it gives only the angles between the pairs of roots in the base, without indicating which is the longer. Two mutually inverse systems (like B_n and C_n; cf. Sec. 16) have the same Coxeter graph.

However, the Cartan matrix is determined if we specify the *ratios of lengths* of the roots. This leads us to attach, to the vertices of the Coxeter graph, coefficients proportional to the square (α, α) of the length of the relevant root α. The Coxeter graph, thus labelled, is called the *Dynkin diagram* of R.

If we agree to identify two Dynkin diagrams which differ only by a coefficient of proportionality, we have:

Proposition 13. *Specifying a Dynkin diagram is equivalent to specifying a Cartan matrix. They determine the root system up to isomorphism.*

Let us explain how to determine the Cartan matrix from the Dynkin diagram:

if $\alpha = \beta$, we have $n(\alpha, \beta) = 2$;

if $\alpha \neq \beta$, and if α and β are not joined by an edge, we have $n(\alpha, \beta) = 0$;

if $\alpha \neq \beta$, if α and β are joined by at least one edge, and if the coefficient of α is less than or equal to that of β, we have $n(\alpha, \beta) = -1$;

if $\alpha \neq \beta$, if α and β are joined by i edges $(1 \leqslant i \leqslant 3)$, and if the coefficient of α is greater than or equal to that of β, we have $n(\alpha, \beta) = -i$.

(In this last case, the coefficient of α is i times that of β; hence there is no need to draw multiple edges.)

Theorem* 4. *Each nonempty connected Dynkin diagram is isomorphic to one of the following:*

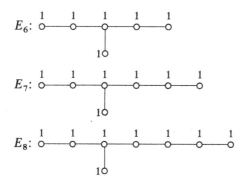

This follows easily from Theorem 3.

Remarks. (1) Conversely, the Dynkin diagrams $A_n, \ldots, E_8$ do indeed correspond to root systems. This can be seen by constructing these systems explicitly; we shall do this in the next section.

(2) It follows from Prop. 13 that the automorphism group E of the Coxeter matrix (cf. Sec. 11) is isomorphic to that of the Dynkin diagram. A quick glance at the list in Theorem 4 shows that:

$E = \{1\}$ for types $A_1, B_n, C_n, G_2, F_4, E_7, E_8$.
E is a group with two elements for $A_n (n \geq 2)$, $D_n (n \geq 5)$, and E_6.
E is isomorphic to the group of permutations of three symbols for type D_4.

(3) A Dynkin diagram is often represented by a symbol such as

$$B_n: \circ\!\!-\!\!-\!\!-\!\!\circ\!\!-\!\!-\quad\cdots\quad-\!\!-\!\!\circ\!\!\Rightarrow\!\!\circ$$

where the inequality sign $>$ on the multiple edge indicates which of the two adjacent roots is the longer (and the absence of an inequality sign means that two adjacent roots have the same length). With this notation, we have

$$C_n: \circ\!\!-\!\!-\!\!-\!\!\circ\!\!-\!\!-\quad\cdots\quad-\!\!-\!\!\circ\!\!\Leftarrow\!\!\circ$$

$$G_2: \circ\!\!\Rightarrow\!\!\circ$$

$$F_4: \circ\!\!-\!\!-\!\!-\!\!\circ\!\!\Rightarrow\!\!\circ\!\!-\!\!-\!\!-\!\!\circ$$

16. Construction of Irreducible Root Systems

In this section, we let $e_1, \ldots, e_n$ denote the standard basis of $\mathbf{R}^n$, and we put on $\mathbf{R}^n$ the bilinear form $(,)$ given by $(e_i, e_j) = \delta_{ij}$. We let L_n denote the subgroup generated by the vectors e_i.

Construction of $A_n (n \geq 1)$. We take V to be the hyperplane of $\mathbf{R}^{n+1}$ orthogonal to $e_1 + \cdots + e_{n+1}$. R consists of elements $\alpha \in V \cap L_{n+1}$ such that $(\alpha, \alpha) = 2$. The symmetry s_α associated with such an element can be written as

$$\beta \mapsto \beta - (\alpha, \beta)\alpha.$$

The fact that R is a root system is immediate.

The elements of R are the vectors $e_i - e_j$, $i \neq j$. For a base S we can take the set of all vectors $e_i - e_{i+1}$, $1 \leq i \leq n$.

The Weyl group can be identified with the group of permutations of $e_1, \ldots, e_{n+1}$.

Construction of B_n ($n \geq 1$). In the space $V = \mathbf{R}^n$, we take R to be the set of all $\alpha \in L_n$ such that $(\alpha, \alpha) = 1$ or $(\alpha, \alpha) = 2$. These are the vectors e_i and $\pm e_i \pm e_j$ $(i \neq j)$.

Base: $e_1 - e_2, e_2 - e_3, \ldots, e_{n-1} - e_n, e_n$.

Weyl group: permutations and sign changes of the vectors e_i.

(For $n = 1$, this system is isomorphic to A_1.)

Construction of C_n ($n \geq 1$). We take the inverse of B_n; it consists of the vectors $\pm e_i \pm e_j$ $(i \neq j)$ and $\pm 2e_i$.

Base: $e_1 - e_2, e_2 - e_3, \ldots, e_{n-1} - e_n, 2e_n$.

Weyl group: the same as that of B_n.

(For $n = 1$, this system is isomorphic to A_1 and B_1; for $n = 2$ it is isomorphic to B_2.)

Construction of D_n ($n \geq 2$). $V = \mathbf{R}^n$; R is the set of all $\alpha \in L_n$ such that $(\alpha, \alpha) = 2$; these are the vectors $\pm e_i \pm e_j$ $(i \neq j)$.

Base: $e_1 - e_2, e_2 - e_3, \ldots, e_{n-1} - e_n, e_{n-1} + e_n$.

Weyl group: permutations and sign changes of (an even number of) the vectors e_i.

(For $n = 2$, this system is isomorphic to $A_1 \times A_1$; for $n = 3$ it is isomorphic to A_3.)

Construction of G_2. This was done in Sec. 3. To put it briefly, R can be described as the set of integers of norm 1 or 3 in the field of the cubic roots of unity.

Construction of F_4. In $V = \mathbf{R}^4$, let L_4' be the subgroup generated by L_4 and $\frac{1}{2}(e_1 + e_2 + e_3 + e_4)$. We take R to be the set of all $\alpha \in L_4'$ such that $(\alpha, \alpha) = 1$ or $(\alpha, \alpha) = 2$. These are the vectors $\pm e_i$, $\pm e_i \pm e_j$ $(i \neq j)$, and $\frac{1}{2}(\pm e_1 \pm e_2 \pm e_3 \pm e_4)$.

Base: $e_2 - e_3, e_3 - e_4, e_4, \frac{1}{2}(e_1 - e_2 - e_3 - e_4)$.

Construction of E_8. In $V = \mathbf{R}^8$, let L'_8 be the subgroup generated by L_8 and $\frac{1}{2}(e_1 + \cdots + e_8)$, and let L''_8 be the subgroup of L'_8 formed by the elements whose sum of coordinates is an *even* integer. We take R to be the set of all $\alpha \in L''_8$ such that $(\alpha, \alpha) = 2$. These are the vectors:

$$\pm e_i \pm e_j \, (i \neq j) \quad \text{and} \quad \frac{1}{2} \sum_{i=1}^{i=8} (-1)^{m(i)} e_i \quad \text{with} \sum m(i) \text{ even.}$$

Construction of E_6 and E_7. We take the intersection of the system E_8 constructed above with the vector subspace spanned by the first six (resp. first seven) elements of the base.

Nonreduced Root Systems. One can show that, for each $n \geqslant 1$ there is (up to isomorphism) exactly one nonreduced irreducible root system: it is the system BC_n obtained as the *union* of the systems B_n and C_n constructed above.

17. Complex Root Systems

Let V be a finite-dimensional *complex* vector space. The definition of a symmetry given in Sec. 1 can be used without change, and Lemma 1 is still true. Hence we have the concept of a root system:

Definition 7. *A subset R of V is called a (complex) root system if*:

(1) *R is finite, spans V (as a complex vector space), and does not contain 0.*
(2) *For each $\alpha \in R$, there is a symmetry $s_\alpha = 1 - \alpha^* \otimes \alpha$ with vector α which leaves R invariant.*
(3) *If $\alpha, \beta \in R$, $s_\alpha(\beta) - \beta$ is an integer multiple of α.*

EXAMPLE. Let R be a root system in a *real* vector space V_0, and let V be the *complexification* $V_0 \otimes \mathbf{C}$ of V_0. The space V_0 is imbedded in V, and R *is a root system in V.* This can be seen by extending the symmetries s^0_α of V_0 by linearity to V.

Theorem 5. *Every complex root system can be obtained in the above way.*

More precisely:

Theorem 5'. *Let R be a root system in a complex vector space V. Let V_0 be the $\mathbf{R}$-subspace of V spanned by R. Then*:

(a) *R is a root system in V_0.*
(b) *The canonical mapping $i: V_0 \otimes \mathbf{C} \to V$ is an isomorphism.*
(c) *If $\alpha \in R$, the symmetry s_α of V is the linear extension of the symmetry s^0_α of V_0.*

PROOF OF (a). Clearly R spans V_0. Furthermore, if $\alpha \in R$, the symmetry s_α leaves invariant R, and hence V_0. Let s_α^0 be its restriction to V_0. If $\beta \in R$, we have $s_\alpha^0(\beta) = \beta - \alpha^*(\beta)\alpha$, with $\alpha^*(\beta) \in \mathbf{Z}$. Hence R is a root system in V_0; moreover, the inverse root α_0^* of α in V_0^* is none other than the image of $\alpha^* \in V^*$ under the restriction homomorphism $V^* \to \mathrm{Hom}(V_0, \mathbf{C})$. $\qquad\square$

PROOF OF (b). Because R spans V, the homomorphism

$$i: V_0 \otimes \mathbf{C} \to V$$

is *surjective*. On the other hand, we have just seen that its transpose

$$t_{i: V^* \to V_0^* \otimes \mathbf{C}}$$

maps α^* to α_0^* for each $\alpha \in R$. But, by Prop. 2, the elements α_0^* form a root system in V_0^*, and in particular they span V_0^*. It follows that t_i is *surjective*, and hence i is injective, giving (b). $\qquad\square$

Finally, (c) follows from the facts proved above.

Theorem 5 reduces the theory of complex root systems to that of real root systems. All the definitions and results of the preceding sections are therefore applicable in the complex case.

CHAPTER VI

Structure of Semisimple Lie Algebras

Throughout this chapter, $\mathfrak{g}$ denotes a *complex semisimple Lie algebra*, and $\mathfrak{h}$ a *Cartan subalgebra* of $\mathfrak{g}$ (cf. Chap. III).

1. Decomposition of $\mathfrak{g}$

If α is an element of the dual space $\mathfrak{h}^*$ of $\mathfrak{h}$, we let $\mathfrak{g}^\alpha$ denote the corresponding eigen-subspace of $\mathfrak{g}$, or in other words, the set of $x \in \mathfrak{g}$ such that $[H, x] = \alpha(H)x$ for all $H \in \mathfrak{h}$; an element of $\mathfrak{g}^\alpha$ is said to have *weight* α.

In particular, $\mathfrak{g}^0$ is the set of elements $x \in \mathfrak{g}$ commuting with $\mathfrak{h}$; by Theorem 3 of Chap. III, we have

$$\mathfrak{g}^0 = \mathfrak{h}.$$

Any element α of $\mathfrak{h}^*$ such that $\alpha \neq 0$ and $\mathfrak{g}^\alpha \neq 0$ is called a *root* of $\mathfrak{g}$ (relative to $\mathfrak{h}$); the set of roots will be denoted by R.

Theorem 1. *One has* $\mathfrak{g} = \mathfrak{h} \oplus \sum_{\alpha \in R} \mathfrak{g}^\alpha$ *(direct sum).*

By Theorem 3 of Chap. III, the endomorphisms $\mathrm{ad}(H)$ of $\mathfrak{g}$ for $H \in \mathfrak{h}$ are diagonalizable; since they commute with each other, they are simultaneously diagonalizable, which proves the theorem.

The subspaces $\mathfrak{g}^\alpha$ have the following properties:

Theorem 2. (a) *R is a root system in $\mathfrak{h}^*$, in the sense of Sec. V.17; this system is reduced* (V.2).
(b) *Let $\alpha \in R$. Then $\mathfrak{g}^\alpha$ is one dimensional; so is the subspace $\mathfrak{h}_\alpha = [\mathfrak{g}^\alpha, \mathfrak{g}^{-\alpha}]$*

of $\mathfrak{h}$. *There is a unique element* $H_\alpha \in \mathfrak{h}_\alpha$ *such* $\alpha(H_\alpha) = 2$; *it is the inverse root of* α (cf. Sec. V.2).

(c) *Let* $\alpha \in R$. *For each nonzero element* X_α *of* $\mathfrak{g}^\alpha$, *there is a unique element* Y_α *of* $\mathfrak{g}^{-\alpha}$ *such that* $[X_\alpha, Y_\alpha] = H_\alpha$. *One has* $[H_\alpha, X_\alpha] = 2X_\alpha$ *and* $[H_\alpha, Y_\alpha] = -2Y_\alpha$. *The subalgebra* $\mathfrak{s}_\alpha = \mathfrak{h}_\alpha \oplus \mathfrak{g}^\alpha \oplus \mathfrak{g}^{-\alpha}$ *is isomorphic to* $\mathfrak{sl}_2$.

(d) *If* $\alpha, \beta \in R$ *and* $\alpha + \beta \neq 0$, *then*

$$[\mathfrak{g}^\alpha, \mathfrak{g}^\beta] = \mathfrak{g}^{\alpha+\beta}.$$

The proof will be given in Sec. 2. In the following theorem, we let (,) denote an invariant nondegenerate symmetric bilinear form on $\mathfrak{g}$ (for example, the Killing form).

Theorem 3. (i) *The subspaces* $\mathfrak{g}^\alpha$ *and* $\mathfrak{g}^\beta$ *are orthogonal if* $\alpha + \beta \neq 0$. *The subspaces* $\mathfrak{g}^\alpha$ *and* $\mathfrak{g}^{-\alpha}$ *are dual with respect to* (,). *The restriction of* (,) *to* $\mathfrak{h}$ *is nondegenerate.*

(ii) *If* $x \in \mathfrak{g}^\alpha$, $y \in \mathfrak{g}^{-\alpha}$, *and* $H \in \mathfrak{h}$, *then*

$$(H, [x, y]) = \alpha(H) \cdot (x, y).$$

(iii) *Let* $\alpha \in R$, *and let* h_α *be the element of* $\mathfrak{h}$ *corresponding to* α *under the isomorphism* $\mathfrak{h} \to \mathfrak{h}^*$ *associated with the chosen bilinear form. Then*

$$[x, y] = (x, y)h_\alpha \qquad \text{if } x \in \mathfrak{g}^\alpha, y \in \mathfrak{g}^{-\alpha}.$$

PROOF. (i) If $x \in \mathfrak{g}^\alpha$, $y \in \mathfrak{g}^\beta$, and $H \in \mathfrak{h}$, we have

$$([H, x], y) + (x, [H, y]) = 0$$

since (,) is invariant. We can also write this as

$$\alpha(H) \cdot (x, y) + \beta(H) \cdot (x, y) = 0.$$

If $\alpha + \beta \neq 0$, we can choose H so that $\alpha(H) + \beta(H) \neq 0$, from which we obtain $(x, y) = 0$, which indeed proves that $\mathfrak{g}^\alpha$ and $\mathfrak{g}^\beta$ are orthogonal.

The decomposition

$$\mathfrak{g} = \mathfrak{h} \oplus \sum (\mathfrak{g}^\alpha + \mathfrak{g}^{-\alpha})$$

is therefore a decomposition of $\mathfrak{g}$ into mutually orthogonal subspaces. Since (,) is nondegenerate, its restriction to each of these subspaces is also nondegenerate; the last two assertions of (i) follow from this.

(ii) The invariance of (,) allows us to write

$$(H, [x, y]) = ([H, x], y) = \alpha(H) \cdot (x, y).$$

(iii) If $H \in \mathfrak{h}$, we have $\alpha(H) = (H, h_\alpha)$ by definition of h_α. Formula (ii) can then be written as

$$(H, [x, y]) = (H, (x, y) \cdot h_\alpha).$$

Since the restriction of (,) to $\mathfrak{h}$ is nondegenerate, we deduce from this that $[x, y] = (x, y) \cdot h_\alpha$.

2. Proof of Theorem 2

This rests basically on Theorem 3 and the properties of the algebra $\mathfrak{sl}_2$ proved in Chap. IV. One proceeds by stages:

2.1. *If $\alpha, \beta \in \mathfrak{h}^*$, then $[\mathfrak{g}^\alpha, \mathfrak{g}^\beta] \subset \mathfrak{g}^{\alpha+\beta}$.* This follows from the Jacobi identity

$$[H, [x, y]] = [[H, x], y] + [x, [H, y]]$$

applied to $H \in \mathfrak{h}$, $x \in \mathfrak{g}^\alpha$, $y \in \mathfrak{g}^\beta$.

2.2. *R spans $\mathfrak{h}^*$.* Otherwise, there would be a nonzero element $H \in \mathfrak{h}$ such that $\alpha(H) = 0$ for all $\alpha \in R$. We would then have $\mathrm{ad}(H) = 0$, that is, H would belong to the *center* of $\mathfrak{g}$; however, we know that the center of $\mathfrak{g}$ is trivial.

2.3. *If $\alpha \in R$, the subspace $\mathfrak{h}_\alpha = [\mathfrak{g}^\alpha, \mathfrak{g}^{-\alpha}]$ of $\mathfrak{h}$ is 1-dimensional.* Indeed, by Theorem 3 (iii), $\mathfrak{h}_\alpha$ consists of the multiples of the element h_α.

2.4. *If $\alpha \in R$, there is a unique element H_α of $\mathfrak{h}_\alpha$ such that $\alpha(H_\alpha) = 2$.* In view of 2.3, it is sufficient to prove that the restriction of α to $\mathfrak{h}_\alpha$ is nontrivial. Let us suppose that it is trivial, and let us choose elements x, y of $\mathfrak{g}^\alpha$, $\mathfrak{g}^{-\alpha}$ such that $z = [x, y]$ is nonzero. Since $\alpha(z) = 0$, we have

$$[z, x] = 0, \quad [z, y] = 0, \quad [x, y] = z.$$

These formulae show that the subalgebra $\mathfrak{a}$ of $\mathfrak{g}$ generated by x, y, z is solvable (and even nilpotent). If $\rho \colon \mathfrak{a} \to \mathrm{End}(V)$ is a finite-dimensional linear representation of $\mathfrak{a}$, Lie's Theorem (Chap. I, Theorem 3) shows that there is a flag D of V stable under $\rho(\mathfrak{a})$. Since $z \in [\mathfrak{a}, \mathfrak{a}]$, we have $\rho(z) \in \mathfrak{n}(D)$, and hence $\rho(z)$ is nilpotent. By applying this result to the representation $\mathrm{ad} \colon \mathfrak{a} \to \mathrm{End}(\mathfrak{g})$, we see that z is nilpotent. On the other hand, z is semisimple (Chap. III, Theorem 3); hence $z = 0$, a contradiction.

2.5. *Let $\alpha \in R$ and let X_α be a nonzero element of $\mathfrak{g}^\alpha$. There exists $Y_\alpha \in \mathfrak{g}^{-\alpha}$ such that $[X_\alpha, Y_\alpha] = H_\alpha$.* Indeed, since $\mathfrak{g}^\alpha$ and $\mathfrak{g}^{-\alpha}$ are dual with respect to $(\ ,\ )$, there exists $y \in \mathfrak{g}^{-\alpha}$ such that $(X_\alpha, y) \neq 0$, giving $[X_\alpha, y] \neq 0$ by Theorem 3 (iii). Multiplying y by a suitable scalar, we obtain the required element Y_α. We have

$$[H_\alpha, X_\alpha] = \alpha(H_\alpha) X_\alpha = 2X_\alpha, \quad [H_\alpha, Y_\alpha] = -\alpha(H_\alpha) Y_\alpha = -2Y_\alpha.$$

If $\mathfrak{s}_\alpha$ is the subalgebra of $\mathfrak{g}$ generated by X_α, Y_α, and H_α, the mapping $(X, Y, H) \mapsto (X_\alpha, Y_\alpha, H_\alpha)$ defines an isomorphism ϕ_α from $\mathfrak{sl}_2$ onto $\mathfrak{s}_\alpha$. Using the adjoint representation, we can now regard $\mathfrak{g}$ as an $\mathfrak{sl}_2$-module.

2.6. *We have $\dim \mathfrak{g}^\alpha = 1$ if $\alpha \in R$.* Let us keep the notation of 2.5. Since $\mathfrak{g}^\alpha$ and $\mathfrak{g}^{-\alpha}$ are dual with respect to $(\ ,\ )$, if we had $\dim \mathfrak{g}^\alpha \neq 1$ there would be a nonzero element $y \in \mathfrak{g}^{-\alpha}$ orthogonal to X_α. By Theorem 3, we would have $[X_\alpha, y] = 0$ and moreover $[H_\alpha, y] = -\alpha(H_\alpha) y = -2y$. Thus y would be a primitive element of weight -2 in $\mathfrak{g}$, regarded as an $\mathfrak{sl}_2$-module by means of ϕ_α. This contradicts Corollary 2 to Theorem 1 of Chap. IV.

2.7. *We have* $\mathfrak{s}_\alpha = \mathfrak{h}_\alpha \oplus \mathfrak{g}^\alpha \oplus \mathfrak{g}^{-\alpha}$. This follows from 2.5 and 2.6.

2.8. *The element* Y_α *in 2.5 is unique.* This follows from the fact that $\dim \mathfrak{g}^{-\alpha} = 1$.

2.9. *If α and β are roots, then $\beta(H_\alpha)$ is an integer and $\beta - \beta(H_\alpha)\alpha$ is a root.* Let $y \in \mathfrak{g}^\beta$, $y \neq 0$, and let $p = \beta(H_\alpha)$. We have

$$[H_\alpha, y] = \beta(H_\alpha)y = p \cdot y,$$

which shows that y has "weight" p, when we view $\mathfrak{g}$ as an $\mathfrak{sl}_2$-module by means of ϕ_α. By Theorem 4 of Chap. IV it follows that p is an integer. We put

$$z = Y_\alpha^p y \quad \text{if } p \geqslant 0, \quad \text{and} \quad z = X_\alpha^{-p} y \quad \text{if } p \leqslant 0.$$

The same theorem shows that $z \neq 0$; since z has weight $\beta - p\alpha$, it follows that $\beta - p\alpha$ is a root.

2.10. *R is a root system, and H_α is the inverse root of α.* By 2.2, R spans $\mathfrak{h}^*$. Now if $\alpha \in R$, let s_α be the endomorphism $\beta \mapsto \beta - \beta(H_\alpha)\alpha$. Since $\alpha(H_\alpha) = 2$, s_α is a symmetry with respect to α (cf. Chap. V). By 2.9, s_α leaves R invariant, and $\beta(H_\alpha)$ is an integer for each $\beta \in R$; the statement now follows.

2.11. *The root system R is reduced.* Otherwise, there would be some $\alpha \in R$ such that $2\alpha \in R$. Let y be a nonzero element of $\mathfrak{g}^{2\alpha}$. We have $[H_\alpha, y] = 2\alpha(H_\alpha)y = 4y$. On the other hand, 3α is not a root, so $\mathrm{ad}(X_\alpha)y = 0$. The formula $H_\alpha = [X_\alpha, Y_\alpha]$ shows that $\mathrm{ad}(H_\alpha)y = \mathrm{ad}(X_\alpha)\,\mathrm{ad}(Y_\alpha)y$; but $\mathrm{ad}(Y_\alpha)y$ belongs to $\mathfrak{g}^\alpha$, so it is a multiple of X_α and is annihilated by $\mathrm{ad}(X_\alpha)$. Thus $4y = \mathrm{ad}(H_\alpha)y = 0$, a contradiction.

2.12. *Let α and β be two nonproportional roots. Let p (resp. q) be the greatest integer such that $\beta - p\alpha$ (resp. $\beta + q\alpha$) is a root. Let $E = \sum_k \mathfrak{g}^{\beta + k\alpha}$. Then E is an irreducible $\mathfrak{s}_\alpha$-module, of dimension $p + q + 1$. If $-p \leqslant k \leqslant q - 1$, the map*

$$\mathrm{ad}(X_\alpha) \colon \mathfrak{g}^{\beta + k\alpha} \to \mathfrak{g}^{\beta + (k+1)\alpha}$$

is an isomorphism. We have $\beta(H_\alpha) = p - q$.

Clearly E is an $\mathfrak{s}_\alpha$-submodule of $\mathfrak{g}$; moreover, if we view E as an $\mathfrak{sl}_2$-module (using ϕ_α to identify $\mathfrak{sl}_2$ and $\mathfrak{s}_\alpha$), we see that the weights of E are the integers $\beta(H_\alpha) + 2k$, for all values of k such that $\beta + k\alpha$ is a root; each such weight has multiplicity 1. These properties of the weights of E, together with the structure theorem of Sec. IV.5, imply that E is irreducible of dimension $m + 1$, with $m = \beta(H_\alpha) + 2q = -\beta(H_\alpha) + 2p$. The fact that

$$\mathrm{ad}(X_\alpha) \colon \mathfrak{g}^{\beta + k\alpha} \to \mathfrak{g}^{\beta + (k+1)\alpha}, \qquad -p \leqslant k \leqslant q - 1$$

is an isomorphism follows from the structure of the irreducible $\mathfrak{sl}_2$-modules.

2.13. *If* $\alpha \in R$, $\beta \in R$, $\alpha + \beta \in R$, *we have* $[\mathfrak{g}^\alpha, \mathfrak{g}^\beta] = \mathfrak{g}^{\alpha+\beta}$. With the notation above we have $q \geqslant 1$. Taking $k = 0$, we deduce that

$$\mathrm{ad}(X_\alpha): \mathfrak{g}^\beta \to \mathfrak{g}^{\beta+\alpha}$$

is an isomorphism, thus proving the result.

The proof of Theorem 2 is now complete.

Remark. Let W be the *Weyl group* associated with the root system R; let us identify W with a group of automorphisms of $\mathfrak{h}$. Then *every element* $w \in W$ *is induced by an inner automorphism of* $\mathfrak{g}$ *which leaves* $\mathfrak{h}$ *invariant*. In fact, it is sufficient to see this when w is the symmetry associated with the root α, and in this case we can take the inner automorphism to be the element

$$\theta_\alpha = e^{\mathrm{ad}(X_\alpha)} e^{-\mathrm{ad}(Y_\alpha)} e^{\mathrm{ad}(X_\alpha)} \quad \text{(cf. Sec. IV.5)}.$$

Conversely, one can show that every inner automorphism of $\mathfrak{g}$ leaving $\mathfrak{h}$ invariant induces an element of W on $\mathfrak{h}$. Moreover, the group $\mathrm{Aut}(\mathfrak{g})/\mathrm{Aut}^0(\mathfrak{g})$ of "outer automorphisms" of $\mathfrak{g}$ can be identified with the group $E = \mathrm{Aut}(R)/W$ defined in Sec. V.11. See *Séminaire S. Lie*, exposé 16, or Bourbaki, Chap. 8, Sec. 5.

3. Borel Subalgebras

Let R be the root system associated with $(\mathfrak{g}, \mathfrak{h})$ and let us choose a *base* S of R (cf. Sec. V.8). Let R^+ be the set of positive roots with respect to S. We put

$$\mathfrak{n} = \sum_{\alpha > 0} \mathfrak{g}^\alpha, \quad \mathfrak{n}^- = \sum_{\alpha > 0} \mathfrak{g}^{-\alpha}, \quad \mathfrak{b} = \mathfrak{h} \oplus \mathfrak{n}.$$

Theorem 4. (a) *One has* $\mathfrak{g} = \mathfrak{n}^- \oplus \mathfrak{h} \oplus \mathfrak{n} = \mathfrak{n}^- \oplus \mathfrak{b}$.

(b) $\mathfrak{n}$ *and* $\mathfrak{n}^-$ *are subalgebras of* $\mathfrak{g}$ *consisting of nilpotent elements; they are nilpotent.*

(c) $\mathfrak{b}$ *is a solvable subalgebra of* $\mathfrak{g}$; *its derived algebra is* $\mathfrak{n}$.

(a) is trivial.

(b): Let $x \in \mathfrak{n}$; for each integer $k > 0$, and each $\beta \in \mathfrak{h}^*$, one has

$$\mathrm{ad}(x)^k(\mathfrak{g}^\beta) \subset \sum_{\alpha_i > 0} \mathfrak{g}^{\beta + \alpha_1 + \cdots + \alpha_k}.$$

If k is sufficiently large, β and $\beta + \alpha_1 + \cdots + \alpha_k$ cannot both be in $R \cup \{0\}$. Hence one has $\mathrm{ad}(x)^k = 0$, which shows that x is nilpotent. The fact that $\mathfrak{n}$ is nilpotent follows from Engel's Theorem (Sec. I.4) or from a direct argument.

The case of $\mathfrak{n}^-$ is similar.

(c) follows from the equation $[\mathfrak{h}, \mathfrak{h}] = \mathfrak{n}$.

The algebra $\mathfrak{b}$ is called the *Borel subalgebra* corresponding to $\mathfrak{h}$ and S.

Theorem* 5 (Borel–Morozov). *Every solvable subalgebra of* $\mathfrak{g}$ *can be mapped to a subalgebra of* $\mathfrak{h}$ *by an inner automorphism of* $\mathfrak{g}$. *In particular,* $\mathfrak{b}$ *is a maximal solvable subalgebra of* $\mathfrak{g}$.

For the proof see A. Borel, *Ann. of Maths.* **64** (1956), pp. 66–67; Bourbaki, Chap. 8, Sec. 10; or Humphreys, Sec. 16.3.

Corollary. *Every subalgebra of* $\mathfrak{g}$ *consisting of nilpotent elements can be mapped to a subalgebra of* $\mathfrak{n}$ *by an inner automorphism of* $\mathfrak{g}$.

This follows from Theorem 5, and from the fact that each nilpotent element of $\mathfrak{g}$ contained in $\mathfrak{b}$ belongs to $\mathfrak{n}$.

4. Weyl Bases

We will keep the notation of the preceding section. We let $(\alpha_1, \ldots, \alpha_n)$ denote the chosen base S; $n = \dim \mathfrak{h}$ is the *rank* of $\mathfrak{g}$ (cf. Chap. III). For each i, we put $H_i = H_{\alpha_i}$, and we choose elements $X_i \in \mathfrak{g}^{\alpha_i}$, $Y_i \in \mathfrak{g}^{-\alpha_i}$ such that $[X_i, Y_i] = H_i$ (cf. Theorem 2).

Finally, we put

$$n(i,j) = \alpha_j(H_i).$$

The matrix formed by the numbers $n(i,j)$ is the *Cartan matrix* of the given system; recall (cf. Sec. V.11) that $n(i,j)$ is an integer $\leqslant 0$ if $i \neq j$.

Theorem 6. (a) $\mathfrak{n}$ *is generated by the elements* X_i, $\mathfrak{n}^-$ *by the elements* Y_i, *and* $\mathfrak{g}$ *by the elements* X_i, Y_i, H_i.

(b) *These elements satisfy the relations* (called the "Weyl relations")

$$[H_i, H_j] = 0,$$

$$[X_i, Y_i] = H_i, \qquad\qquad [X_i, Y_j] = 0 \quad \text{if } i \neq j,$$

$$[H_i, X_j] = n(i,j)X_j, \qquad [H_i, Y_j] = -n(i,j)Y_j.$$

(c) *They also satisfy the following relations*:

$$(\theta_{ij}) \quad \operatorname{ad}(X_i)^{-n(i,j)+1}(X_j) = 0 \qquad (i \neq j),$$

$$(\theta_{ij}^-) \quad \operatorname{ad}(Y_i)^{-n(i,j)+1}(Y_j) = 0 \qquad (i \neq j).$$

PROOF. (a) It is sufficient to show that $\mathfrak{n}$ is generated by the elements X_i. To prove this, let $\alpha \in R^+$. It is known (cf. Sec. V.9) that α can be decomposed as a sum of roots α_i,

$$\alpha = \alpha_{i_1} + \cdots + \alpha_{i_k},$$

in such a way that the partial sums $\alpha_{i_1} + \cdots + \alpha_{i_h}$ belong to R^+ for each $h \leqslant k$. Let us choose such a decomposition and put

$$X_\alpha = [X_{i_k}, [X_{i_{k-1}}, \ldots, [X_{i_2}, X_{i_1}] \ldots]].$$

By Theorem 2, X_α is a nonzero element of $\mathfrak{g}^\alpha$. Since $\mathfrak{n}$ is the sum of the spaces $\mathfrak{g}^\alpha$, $\alpha \in R^+$, this shows that $\mathfrak{n}$ is indeed generated by the elements X_i.

(b) The relation $[X_i, Y_j] = 0$ for $i \neq j$ follows from the fact that $[X_i, Y_j]$ has weight $\alpha_i - \alpha_j$, where we know that $\alpha_i - \alpha_j$ is not a root (because every root is a linear combination of roots α_i with coefficients *of the same sign*). The other relations are obvious.

(c) The element

$$\theta_{ij} = \text{ad}(X_i)^{-n(i,j)+1}(X_j)$$

has weight $\alpha_j - n(i,j)\alpha_i + \alpha_i = s_i(\alpha_j - \alpha_i)$, where s_i denotes the symmetry corresponding to α_i. Since $\alpha_j - \alpha_i$ is not a root, neither is $s_i(\alpha_j - \alpha_i)$, so $\theta_{ij} = 0$. The relation $\theta_{ij}^- = 0$ is proved in the same way. $\square$

Theorem 7. (i) *The algebra $\mathfrak{n}$ can be defined by the generators X_i and the relations θ_{ij}, $i \neq j$.*

(ii) *The algebra $\mathfrak{g}$ can be defined by the generators X_i, Y_i, H_i, the Weyl relations, and the relations θ_{ij}, θ_{ij}^-.*

((i) shows that the generators X_i and the relations θ_{ij} form a "presentation" of $\mathfrak{n}$; in other words, if L is the free Lie algebra generated by the elements X_i, and if $f: L \rightarrow \mathfrak{n}$ is the obvious homomorphism, then the kernel of f is the ideal of L generated by the elements θ_{ij}. The meaning of (ii) is similar.)

For the proof, see the Appendix at the end of this chapter.

EXAMPLE. If R is of type G_2, the algebra $\mathfrak{n}$ has a presentation consisting of two generators X_1, X_2 and of two relations:

$$[X_1, [X_1, X_2]] = 0, \quad [X_2, [X_2, [X_2, [X_2, X_1]]]] = 0.$$

We now give an application of Theorem 7:

Corollary. *There is an automorphism σ of $\mathfrak{g}$ which is equal to -1 on $\mathfrak{h}$, and which sends X_i to $-Y_i$, Y_i to $-X_i$, for all i. One has $\sigma^2 = 1$.*

Let us put $H_i' = -H_i$, $X_i' = -Y_i$, $Y_i' = -X_i$. Clearly the elements X_i', Y_i', H_i' satisfy the Weyl relations and the relations θ_{ij}, θ_{ij}^-. Hence by (ii) there is a homomorphism $\sigma: \mathfrak{g} \rightarrow \mathfrak{g}$ mapping X_i, Y_i, H_i to X_i', Y_i', H_i'. Since σ^2 fixes X_i, Y_i, H_i, it is the identity, as required.

Remark. Theorem 7 gives an explicit description of $\mathfrak{g}$ and of $\mathfrak{n}$ in terms of the *Cartan matrix* $(n(i,j))$.

5. Existence and Uniqueness Theorems

The conjugacy theorem for Cartan subalgebras (Sec. III.4) shows that the root system of a semisimple Lie algebra is *independent* (up to isomorphism) of the chosen Cartan subalgebra. Furthermore:

Theorem 8. *Two semisimple Lie algebras corresponding to isomorphic root systems are isomorphic.*

More precisely:

Theorem 8′. *Let $\mathfrak{g}$ (resp. $\mathfrak{g}'$) be a semisimple Lie algebra, $\mathfrak{h}$ (resp. $\mathfrak{h}'$) a Cartan subalgebra of $\mathfrak{g}$ (resp. $\mathfrak{g}'$), S (resp. S') a base for the corresponding root system, and $r: S \to S'$ a bijection sending the Cartan matrix of S to that of S'. For each $i \in S$ (resp. $j \in S'$), let X_i (resp. X_j') be a nonzero element of $\mathfrak{g}^i$ (resp. $\mathfrak{g}'^j$). Then there is a unique isomorphism $f: \mathfrak{g} \to \mathfrak{g}'$ sending H_i to $H'_{r(i)}$ and X_i to $X'_{r(i)}$ for all $i \in S$.*

Let Y_i (resp. Y_j') be the element of $\mathfrak{g}^{-i}$ (resp. of $\mathfrak{g}'^{-j}$) such that $[X_i, Y_i] = H_i$ (resp. $[X_j', Y_j'] = H_j'$). Then Theorem 7 shows that there is a unique homomorphism $f: \mathfrak{g} \to \mathfrak{g}'$ sending X_i, Y_i, H_i to $X'_{r(i)}$, $Y'_{r(i)}$, $H'_{r(i)}$, and clearly this is an isomorphism.

Remark. By applying this result in the case $\mathfrak{g}' = \mathfrak{g}$, $\mathfrak{h}' = \mathfrak{h}$, $S' = -S$, one obtains another proof of the corollary to Theorem 7.

Finally, here is the existence theorem:

Theorem 9. *Let R be a reduced root system. There exists a semisimple Lie algebra $\mathfrak{g}$ whose root system is isomorphic to R.*

Let $S = \{\alpha_1, \ldots, \alpha_n\}$ be a base for R, with $(n(i,j))$ the corresponding Cartan matrix. Let $\mathfrak{g}$ be the Lie algebra defined by $3n$ generators X_i, Y_i, H_i and by the relations in Theorem 6 (i.e. the Weyl relations and the relations θ_{ij}, θ_{ij}^-). One shows (cf. the Appendix) that this Lie algebra is finite dimensional, semisimple, and has a root system isomorphic to R. Hence the theorem.

Corollary. *For $\mathfrak{g}$ to be simple, it is necessary and sufficient that R should be irreducible.*

This is obvious.

6. Chevalley's Normalization

For each $\alpha \in R$, choose a nonzero element $X_\alpha \in \mathfrak{g}^\alpha$. Then we have

$$[X_\alpha, X_\beta] = \begin{cases} N_{\alpha,\beta} X_{\alpha+\beta} & \text{if } \alpha + \beta \in R, \\ 0 & \text{if } \alpha + \beta \notin R, \quad \alpha + \beta \neq 0, \end{cases}$$

where $N_{\alpha,\beta}$ is a nonzero scalar. The coefficients $N_{\alpha,\beta}$ determine the "multiplication table" of $\mathfrak{g}$. However, they depend on the choice of the elements X_α.

Theorem 10. *One can choose the elements X_α so that*

$$[X_\alpha, X_{-\alpha}] = H_\alpha \qquad \textit{for all } \alpha \in R,$$

$$N_{\alpha,\beta} = -N_{-\alpha,-\beta} \qquad \textit{for } \alpha, \beta, \alpha + \beta \in R.$$

Let R^+ be the set of positive roots relative to a base S of R, and let σ be an automorphism of $\mathfrak{g}$ equal to -1 on $\mathfrak{h}$, and such that $\sigma^2 = 1$ (cf. the Corollary to Theorem 7). We have $\sigma(\mathfrak{g}^\alpha) = \mathfrak{g}^{-\alpha}$. Let $\alpha \in R^+$, and let us choose a nonzero element x_α of $\mathfrak{g}^\alpha$. We have $[x_\alpha, \sigma(x_\alpha)] \in [\mathfrak{g}^\alpha, \mathfrak{g}^{-\alpha}]$, so there is a nonzero scalar t_α such that

$$[x_\alpha, \sigma(x_\alpha)] = t_\alpha H_\alpha.$$

Let u_α be a square root of $-t_\alpha$, and let us put

$$X_\alpha = u_\alpha^{-1} x_\alpha, \quad X_{-\alpha} = -\sigma(X_\alpha).$$

We now have $[X_\alpha, X_{-\alpha}] = H_\alpha$ and $X_\alpha + \sigma(X_\alpha) = 0$. The identity

$$N_{\alpha,\beta} = -N_{-\alpha,-\beta}$$

is then obtained by writing $[\sigma X_\alpha, \sigma X_\beta] = \sigma[X_\alpha, X_\beta]$.

Theorem* 11 (Chevalley). *Suppose that the conditions of Theorem 10 are satisfied. Let α, $\beta \in R$ be such that $\alpha + \beta \in R$, and let p be the greatest integer such that $\beta - p\alpha \in R$ (cf. VI.2.12). Then one has*

$$N_{\alpha,\beta} = \pm(p+1).$$

For the proof, see Chevalley, *Tôkoku Math. J.*, **7** (1955), pp. 22–23, or Bourbaki, Chap. 8, Sec. 2, No. 4.

Remarks. (1) Let $\mathfrak{g}(\mathbf{Z})$ be the abelian subgroup of $\mathfrak{g}$ generated by the elements H_α and by a family of elements X_α satisfying the conditions of Theorem 10. It follows from Theorem 11 that $\mathfrak{g}(\mathbf{Z})$ is a *Lie algebra* over $\mathbf{Z}$. For each field K, one can therefore define the Lie algebra $\mathfrak{g}(K) = \mathfrak{g}(\mathbf{Z}) \otimes K$. This is the starting point for the construction of the "Chevalley groups" (cf. Chevalley, *Tôhoku, loc. cit.*; see also Carter's survey, *J. London Math. Soc.*, **40** (1965), and his book *Simple Groups of Lie Type*, Wiley (1972)).

(2) Tits (*Publ. Math. I.H.E.S.*, **31** (1966), pp. 21–58) has determined the $\pm$ signs in Theorem 11 (however, it is necessary to index the elements X_α differently). From this he has deduced a new proof of the existence theorem (Theorem 9).

(3) Let $\mathfrak{k}$ be the real vector subspace of $\mathfrak{g}$ spanned by the elements iH_α, the elements $X_\alpha - X_{-\alpha}$, and the elements $i(X_\alpha + X_{-\alpha})$. One easily checks that $\mathfrak{k}$ is a real Lie subalgebra of $\mathfrak{g}$, and that the Killing form of $\mathfrak{k}$ is *negative*. Moreover, $\mathfrak{g}$ can be identified with the complexification $\mathfrak{k} \otimes \mathbf{C}$ of $\mathfrak{k}$. One says that $\mathfrak{k}$ is a *compact form* of $\mathfrak{g}$. The existence of such a form is the basis of Weyl's "unitarian trick." When $\mathfrak{g} = \mathfrak{sl}_2$, we have $\mathfrak{k} = \mathfrak{su}_2$ (cf. Secs. IV.6, IV.7).

Appendix. Construction of Semisimple Lie Algebras by Generators and Relations

Let R be a root system in a complex vector space V. For consistency with the notation of the preceding sections, the dual of V will be denoted by $\mathfrak{h}$, so that $V = \mathfrak{h}^*$. Let $S = \{\alpha_1, \ldots, \alpha_n\}$ be a base for R, let $H_1, \ldots, H_n \in \mathfrak{h}$ be the inverse roots of $\alpha_1, \ldots, \alpha_n$, and let

$$n(i,j) = \langle \alpha_j, H_i \rangle.$$

The numbers $n(i,j)$ form the *Cartan matrix* of R with respect to S. We aim to prove the following theorem.

Theorem. *Let $\mathfrak{g}$ be the Lie algebra defined by $3n$ generators X_i, Y_i, H_i and by the relations*

$$(W.1) \quad [H_i, H_j] = 0$$

$$(W.2) \quad [X_i, Y_i] = H_i, \qquad\qquad [X_i, Y_j] = 0 \quad \text{if } i \neq j$$

$$(W.3) \quad [H_i, X_j] = n(i,j)X_j, \qquad [H_i, Y_j] = -n(i,j)Y_j$$

$$(\theta_{ij}) \quad \operatorname{ad}(X_i)^{-n(i,j)+1}(X_j) = 0 \quad \text{if } i \neq j$$

$$(\theta_{ij}^-) \quad \operatorname{ad}(Y_i)^{-n(i,j)+1}(Y_j) = 0 \quad \text{if } i \neq j.$$

Then $\mathfrak{g}$ is a semisimple Lie algebra, with the subalgebra $\mathfrak{h}$ generated by the elements H_i as a Cartan subalgebra; its root system is R.

Let us first consider the algebra $\mathfrak{a}$ defined by the $3n$ generators X_i, Y_i, H_i and by the relations $(W.1), (W.2), (W.3)$. The structure of $\mathfrak{a}$ is known; it has been determined by Chevalley, Harish-Chandra, and Jacobson. We just state the result (for the proof, see, e.g. Bourbaki, Chap. 8, Sec. 4, No. 2):

One has

$$\mathfrak{a} = \mathfrak{n} \oplus \mathfrak{h} \oplus \mathfrak{x},$$

where η (*resp.* $\mathfrak{x}$) *is the Lie algebra generated by the elements* Y_i (*resp.* X_i), *and where* $\mathfrak{h}$ *has the elements* H_i *as a basis. Moreover,* η (*resp.* $\mathfrak{x}$) *can be identified with the free Lie algebra generated by the elements* Y_i (*resp.* X_i).

(By abuse of notation, we identify the original vector space $\mathfrak{h}$ with the vector subspace of $\mathfrak{a}$ spanned by the elements H_i.)

Now let

$$\theta_{ij} = \mathrm{ad}(X_i)^{-n(i,\,j)+1}(X_j)$$

and

$$\theta_{ij}^- = \mathrm{ad}(Y_i)^{-n(i,\,j)+1}(Y_j).$$

We have $\theta_{ij} \in \mathfrak{x}$, $\theta_{ij}^- \in \eta$. We denote by $\mathfrak{u}$ (resp. $\mathfrak{u}^-$) the ideal of $\mathfrak{x}$ (resp. of η) generated by the elements θ_{ij} (resp. θ_{ij}^-) for $i \neq j$. Let $\mathfrak{r} = \mathfrak{u} \oplus \mathfrak{u}^-$.

(a) $\mathfrak{u}$, $\mathfrak{u}^-$, *and* $\mathfrak{r}$ *are ideals of* $\mathfrak{a}$. Let $U\mathfrak{a}$ be the universal enveloping algebra of $\mathfrak{a}$. The adjoint representation $\mathrm{ad} \colon \mathfrak{a} \to \mathrm{End}(\mathfrak{a})$ defines a $U\mathfrak{a}$-module structure on $\mathfrak{a}$. The ideal $\mathfrak{u}_{ij}$ of $\mathfrak{a}$ generated by θ_{ij} is equal to the submodule $(U\mathfrak{a}) \cdot \theta_{ij}$. By the Birkhoff–Witt theorem, $\mathfrak{u}_{ij}$ is spanned (as a vector space) by the elements $XYH\theta_{ij}$, with $X \in U\mathfrak{x}$, $Y \in U\eta$, $H \in U\mathfrak{h}$. Clearly $H\theta_{ij}$ is proportional to θ_{ij}; moreover, a straightforward calculation (cf. Jacobson, Lemma 1, p. 216) shows that $\mathrm{ad}(Y_k)(\theta_{ij}) = 0$ for all k. It follows that $Y\theta_{ij}$ is proportional to θ_{ij}. Thus the ideal $\mathfrak{u}_{ij}$ is generated by the elements $X\theta_{ij}$, and is therefore contained in $\mathfrak{u}$. We then have $\mathfrak{u} = \sum \mathfrak{u}_{ij}$, showing that $\mathfrak{u}$ is indeed an ideal of $\mathfrak{a}$. A similar argument may be applied to $\mathfrak{u}^-$; the result for $\mathfrak{r}$ then follows.

Thus, $\mathfrak{r}$ is the smallest ideal generated by the elements θ_{ij} and θ_{ij}^-. Hence the algebra $\mathfrak{g}$ which we wish to study is simply the quotient $\mathfrak{a}/\mathfrak{r}$.

(b) *One has* $\mathfrak{g} = \mathfrak{n}^- \oplus \mathfrak{h} \oplus \mathfrak{n}$, *where* $\mathfrak{n} = \mathfrak{x}/\mathfrak{u}$, $\mathfrak{n}^- = \eta/\mathfrak{u}^-$. This is obvious.

(c) *The endomorphisms* $\mathrm{ad}(X_i)$ *and* $\mathrm{ad}(Y_i)$ *of* $\mathfrak{g}$ *are locally nilpotent.* Let V_i be the set of all $z \in \mathfrak{g}$ such that $\mathrm{ad}(X_i)^k z = 0$ for some integer k. We must prove that $V_i = \mathfrak{g}$. Now a simple computation shows that V_i is a Lie subalgebra of $\mathfrak{g}$. Since V_i contains the elements X_k (by the relations $\theta_{ij} = 0$) and Y_k (by the relations W), it contains the elements $H_k = [X_k, Y_k]$, so we indeed have $V_i = \mathfrak{g}$. The same argument can be applied to $\mathrm{ad}(Y_i)$.

We now introduce some notation. If λ is a linear form on $\mathfrak{h}$, we denote by $\mathfrak{a}^\lambda$ (resp. $\mathfrak{g}^\lambda$) the set of all $z \in \mathfrak{a}$ (resp. $z \in \mathfrak{g}$) such that $\mathrm{ad}(H)z = \lambda(H)z$ for all $H \in \mathfrak{h}$; such an element z will be said to have *weight* λ. It follows from the decomposition of $\mathfrak{a}$ given above that $\mathfrak{a}$ is the direct sum of the subspaces $\mathfrak{a}^\lambda$; hence $\mathfrak{g}$ is the direct sum of its subspaces $\mathfrak{g}^\lambda$. If $\mathfrak{a}^\lambda \neq 0$, λ is a linear combination of the simple roots α_i, with integer coefficients, all of the same sign.

We have $\mathfrak{h} = \mathfrak{a}^0$, $\mathfrak{x} = \sum_{\lambda > 0} \mathfrak{a}^\lambda$, $\eta = \sum_{\lambda < 0} \mathfrak{a}^\lambda$, and similarly for $\mathfrak{g}$. We will now find the dimension of each $\mathfrak{g}^\lambda$. First, we have:

(d) *If λ is the image of μ under an element w of the Weyl group W, then* $\dim \mathfrak{g}^\lambda = \dim \mathfrak{g}^\mu$. It is sufficient to show this when w is the symmetry corresponding to one of the roots α_i in S (cf. Chap. V, Theorem 2). In this case, we define an automorphism θ_i of $\mathfrak{g}$ by the formula

$$\theta_i = e^{\mathrm{ad}(X_i)} e^{-\mathrm{ad}(Y_i)} e^{\mathrm{ad}(X_i)},$$

which is meaningful because of (c). One easily checks that θ_i induces the symmetry s_i relative to α_i on $\mathfrak{h}$, and hence sends $\mathfrak{g}^\lambda$ to $\mathfrak{g}^\mu$ if $\lambda = s_i(\mu)$. Thus $\dim \mathfrak{g}^\lambda = \dim \mathfrak{g}^\mu$.

(e) *One has* $\dim \mathfrak{g}^{\alpha_i} = 1$ *and* $\dim \mathfrak{g}^{m\alpha_i} = 0$ *for* $m \neq \pm 1, 0$. The corresponding formulae for $\mathfrak{a}$ are obvious, and moreover $\mathfrak{u}$ does not contain X_i. Hence the result.

(f) *If $\alpha \in R$, then* $\dim \mathfrak{g}^\alpha = 1$. Indeed, there exists $w \in W$, sending α to some α_i, so we can use (d) and (e).

(g) *Let λ be a linear combination of the simple roots α_i, with real coefficients, and suppose that λ is not a multiple of any root. Then there exists $w \in W$ such that $w(\lambda) = \sum t_i \alpha_i$, with some $t_i > 0$ and some $t_i < 0$.* Let $\mathfrak{h}_\mathbf{R}$ be the real vector subspace of $\mathfrak{h}$ spanned by the elements H_i (cf. Sec. V.17). We denote by L_α (resp. L) the hyperplane of $\mathfrak{h}_\mathbf{R}$ orthogonal to α (resp. to λ). By hypothesis, L is not equal to any L_α, and hence is not contained in their union. Let H be an element of L which does not lie in any L_α. Transforming λ and L by an element of W if necessary, we can assume that $\alpha_i(H) > 0$ for all i (cf. Chap. V, Theorem 2(a)). If we write λ in the form $\lambda = \sum t_i \alpha_i$, we have

$$0 = \lambda(H) = \sum t_i \alpha_i(H).$$

Since the terms $\alpha_i(H)$ are > 0, this is impossible unless two of the coefficients t_i have opposite signs.

(h) *If λ is not a root, and if $\lambda \neq 0$, then* $\mathfrak{g}^\lambda = 0$. We can assume that λ is a linear combination of the simple roots α_i, with integer coefficients. If λ is a multiple of a root, (h) follows from (d) and (e). Otherwise, by (g) there exists $w \in W$ such that the linear form $\mu = w(\lambda)$ is a linear combination of the roots α_i, with two coefficients of opposite signs. We then have $\mathfrak{a}^\mu = 0$, so that $\mathfrak{g}^\mu = 0$, and we finish by applying (d) again.

(i) *The algebra $\mathfrak{g}$ has finite dimension, equal to* $n + \mathrm{Card}(R)$. Indeed, by (f) and (h) we have

$$\mathfrak{g} = \mathfrak{h} \oplus \sum_{\alpha \in R} \mathfrak{g}^\alpha,$$

and each $\mathfrak{g}$ is one dimensional.

(j) *If* $\alpha \in R$, $[\mathfrak{g}^{\alpha}, \mathfrak{g}^{-\alpha}]$ *consists of multiples of* H_{α}. *The subalgebra* $\mathfrak{s}_{\alpha}$ *of* $\mathfrak{g}$ *generated by* H_{α}, $\mathfrak{g}^{\alpha}$, *and* $\mathfrak{g}^{-\alpha}$ *is isomorphic to* $\mathfrak{sl}_2$. This is clear when α is one of the roots α_i. We can reduce the proof to this case by using a product of automorphisms of the form θ_i.

(k) $\mathfrak{g}$ *is a semisimple algebra*. Let $\mathfrak{t}$ be an abelian ideal of $\mathfrak{g}$. Since $\mathfrak{t}$ is invariant under the endomorphisms $\mathrm{ad}(H)$, $H \in \mathfrak{h}$, we have $\mathfrak{t} = \mathfrak{t} \cap \mathfrak{h} \oplus \sum_{\alpha \in R} \mathfrak{t} \cap \mathfrak{g}^{\alpha}$. However, since $\mathfrak{s}_{\alpha}$ is isomorphic to $\mathfrak{sl}_2$, $\mathfrak{t} \cap \mathfrak{sl}_2 = 0$ and hence *a fortiori* $\mathfrak{t} \cap \mathfrak{g}^{\alpha} = 0$. We therefore have $\mathfrak{t} \subset \mathfrak{h}$. By writing down the condition that $\mathfrak{t}$ is invariant under $\mathrm{ad}(X_i)$, we see that $\mathfrak{t}$ is orthogonal to each α_i; hence $\mathfrak{t} = 0$, and $\mathfrak{g}$ is semisimple.

(l) $\mathfrak{h}$ *is a Cartan subalgebra of* $\mathfrak{g}$, *and* R *is the corresponding root system*. For $\mathfrak{h}$ is equal to its own normalizer. The fact that R is the corresponding root system is clear.

The proof of the theorem is thus complete, and with it that of Theorem 9 (the existence theorem).

Theorem 7 follows easily: let $\mathfrak{g}'$ be a semisimple Lie algebra, $\mathfrak{h}'$ a Cartan subalgebra of $\mathfrak{g}'$, and X_i', Y_i', H_i' the corresponding generators of $\mathfrak{g}'$. Suppose that the Cartan matrix of $\mathfrak{g}'$ is the same as that of the Lie algebra $\mathfrak{g}$ which we have just constructed. By Theorem 6, the generators X_i', Y_i', H_i' satisfy the relations (W) and the relations (θ_{ij}), (θ_{ij}^-); hence there is a homomorphism $f: \mathfrak{g} \to \mathfrak{g}'$ sending X_i, Y_i, H_i to X_i', Y_i', H_i'. Since the latter generate $\mathfrak{g}$, f is surjective. However, $\mathfrak{g}$ and $\mathfrak{g}'$ have the same dimension, namely $n + \mathrm{Card}(R)$; thus f is an isomorphism. Theorem 7 is now obvious.

EXERCISE. Let $\mathfrak{r}'$ be an ideal of $\mathfrak{a}$. Show that the following conditions are equivalent:

(i) $\mathfrak{r}'$ contains the ideal $\mathfrak{r}$ generated by the elements θ_{ij} and θ_{ij}^-.
(ii) The algebra $\mathfrak{a}/\mathfrak{r}'$ is finite dimensional.
(iii) There is an integer $m \geqslant 0$ such that, for all i, j, $i \neq j$, one has $\mathrm{ad}(X_i)^m(X_j) \in \mathfrak{r}'$ and $\mathrm{ad}(Y_i)^m(Y_j) \in \mathfrak{r}'$.

(In particular, $\mathfrak{r}$ is *the smallest ideal of finite codimension in* $\mathfrak{a}$.)

CHAPTER VII

Linear Representations of Semisimple Lie Algebras

In this chapter, $\mathfrak{g}$ denotes a complex semisimple Lie algebra, $\mathfrak{h}$ a Cartan subalgebra of $\mathfrak{g}$, and R the corresponding root system. We choose a base $S = \{\alpha_1, \ldots, \alpha_n\}$ of R, and we denote by R^+ the set of positive roots (with respect to S).

For each $\alpha \in R^+$, we choose $X_\alpha \in \mathfrak{g}^\alpha$, $Y_\alpha \in \mathfrak{g}^{-\alpha}$ so that $[X_\alpha, Y_\alpha] = H_\alpha$ (cf. Chap. VI). When α is one of the simple roots α_i, we write X_i, Y_i, H_i instead of X_{α_i}, Y_{α_i}, H_{α_i}. We put $\mathfrak{n} = \sum_{\alpha > 0} \mathfrak{g}^\alpha$, $\mathfrak{n}^- = \sum_{\alpha < 0} \mathfrak{g}^\alpha$, $\mathfrak{b} = \mathfrak{h} \oplus \mathfrak{n}$.

We intend to study those irreducible $\mathfrak{g}$-modules having a "highest weight" (cf. Sec. 3), and in particular to characterize those which are *finite dimensional*.

1. Weights

Let V be a $\mathfrak{g}$-module (not necessarily finite dimensional), and let $\omega \in \mathfrak{h}^*$ be a linear form on $\mathfrak{h}$. We will let V^ω denote the set of all $v \in V$ such that $Hv = \omega(H)v$ for all $H \in \mathfrak{h}$. This is a vector subspace of V. An element of V^ω is said to *have weight* ω. The dimension of V^ω is called the *multiplicity* of ω in V; if $V^\omega \neq 0$, ω is called a weight of V.

Proposition 1. (a) *One has $\mathfrak{g}^\alpha V^\omega \subset V^{\omega+\alpha}$ if $\omega \in \mathfrak{h}^*$, $\alpha \in R$.*
(b) *The sum $V' = \sum_\omega V^\omega$ is direct; it is a $\mathfrak{g}$-submodule of V.*

Let $X \in \mathfrak{g}^\alpha$, $v \in V^\omega$; if $H \in \mathfrak{h}$, one has

$$H(Xv) = X(Hv) + [H, X]v = (\omega(H) + \alpha(H))Xv,$$

which shows that Xv has weight $\omega + \alpha$, giving (a).

The fact that the sum of the subspaces V^ω is direct is standard (eigenvectors associated with distinct eigenvalues are linearly independent). Finally, (a) shows that V' is invariant under each $\mathfrak{g}^\alpha$, and hence under $\mathfrak{g}$ since $\mathfrak{g} = \mathfrak{h} \oplus \sum \mathfrak{g}^\alpha$.

2. Primitive Elements

Let V be a $\mathfrak{g}$-module, v an element of V, and ω a linear form on $\mathfrak{h}$. One says that v is a *primitive element of weight* ω if it satisfies the following two conditions:

(i) v is nonzero, and has weight ω.
(ii) One has $X_\alpha v = 0$ for all $\alpha \in R^+$ (or for all $\alpha \in S$, which amounts to the same thing).

The primitive elements can also be characterized as the eigenvalues of the Borel subalgebra $\mathfrak{b}$.

Proposition 2. *Let V be a $\mathfrak{g}$-module and let $v \in V$ be a primitive element of V of weight ω; let E be the $\mathfrak{g}$-submodule of V generated by v. Then*:

(1) *If $\beta_1, \ldots, \beta_k$ denote the different positive roots, E is spanned* (as a vector space) *by the elements of the form*

$$Y_{\beta_1}^{m_1} \cdots Y_{\beta_k}^{m_k} v \qquad \text{with } m_i \in \mathbf{N}.$$

(2) *The weights of E have the form*

$$\omega - \sum_{i=1}^{n} p_i \alpha_i, \qquad p_i \in \mathbf{N}.$$

They have finite multiplicity.
(3) *ω is a weight of E of multiplicity 1.*
(4) *E is an indecomposable $\mathfrak{g}$-module.*

(Recall that a module E is called indecomposable if it is nontrivial and if, for each direct sum decomposition $E = E_1 \oplus E_2$, one has $E_1 = 0$ or $E_2 = 0$. Every irreducible module is indecomposable; in general, the converse is false.)

Let $A = U\mathfrak{g}$ be the universal enveloping algebra of $\mathfrak{g}$; similarly let $B = U\mathfrak{b}$ and $C = Un^-$. Since $\mathfrak{g} = \mathfrak{n}^- \oplus \mathfrak{b}$, one has $A = C \cdot B$, and hence $E = A \cdot v = C \cdot B \cdot v$. However, since v is an eigenvector for $\mathfrak{b}$, every product bv, with $b \in B$, is proportional to v; we therefore have $C \cdot B \cdot v = C \cdot v$, that is, $E = C \cdot v$. But by the Birkhoff–Witt Theorem, the monomials

$$Y_{\beta_1}^{m_1} \cdots Y_{\beta_k}^{m_k} \qquad (m_i \in \mathbf{N})$$

form a basis for C; hence (1).

Moreover, Prop. 1 shows that $Y_{\beta_1}^{m_1} \cdots Y_{\beta_k}^{m_k} v$ has weight $\omega - \sum m_j \beta_j$, and each β_j is a linear combination of the simple roots α_i with integer coefficients $\geqslant 0$; hence (2). For (3), we note that $\omega - \sum m_j \beta_j$ cannot be equal to ω unless every m_j is zero.

Finally, if E is the direct sum of two submodules E_1 and E_2, we have $E^\omega = E_1^\omega \oplus E_2^\omega$. Since we have just seen that dim $E^\omega = 1$, we must have $E^\omega = E_1^\omega$ or $E^\omega = E_2^\omega$. In the first case, we have $v \in E_1$, and since v generates E this forces $E = E_1$ and $E_2 = 0$. We apply a similar argument in the second case; thus E is indecomposable.

3. Irreducible Modules with a Highest Weight

Theorem 1. *Let V be an irreducible $\mathfrak{g}$-module containing a primitive element v of weight ω. Then:*

(a) *v is the only primitive element of V (up to scalar multiplication); its weight ω is called the "highest weight" of V.*

(b) *The weights π of V have the form*

$$\pi = \omega - \sum m_i \alpha_i \qquad \text{with } m_i \in \mathbf{N}.$$

They have finite multiplicity; in particular, ω has multiplicity 1. One has $V = \sum V^\pi$.

(c) *For two irreducible $\mathfrak{g}$-modules V_1 and V_2 with highest weights ω_1 and ω_2 to be isomorphic, it is necessary and sufficient that $\omega_1 = \omega_2$.*

(Statement (b) shows that the weights of V are *dominated* by ω, in an obvious sense; this justifies the terminology "highest weight".)

The $\mathfrak{g}$-submodule E of V generated by v is nonzero, and hence equal to V since V is irreducible. By applying Prop. 2 to it, one obtains (b).

Let us now prove (a): let v' be a primitive element of V, of weight ω'. By (b), ω' can be written as

$$\omega' = \omega - \sum m_i \alpha_i, \qquad m_i \geqslant 0.$$

Similarly, exchanging the roles of v and v', we see that

$$\omega = \omega' - \sum m_i' \alpha_i, \qquad m_i' \geqslant 0.$$

These two equations are possible only if $m_i = m_i' = 0$ for all i, that is, $\omega = \omega'$. By (b), v and v' are then proportional giving (a).

For (c), it is sufficient to prove that, if $\omega_1 = \omega_2$, the modules V_1 and V_2 are isomorphic. Let v_i ($i = 1, 2$) be a primitive element of V_i, of weight $\omega = \omega_1 = \omega_2$. Clearly the $\mathfrak{g}$-module $V = V_1 \oplus V_2$ has $v = v_1 + v_2$ as a primitive element of weight ω. Let E be the $\mathfrak{g}$-submodule of V generated by v. The second projection $\text{pr}_2 : V \to V_2$ induces a $\mathfrak{g}$-module homomorphism $f_2 : E \to V_2$. One has $f(v) = v_2$; since v_2 generates V_2, it follows that f is *surjective*. Moreover, the kernel $N_2 = V_1 \cap E$ of f_2 is a submodule of V_1. This submodule does not contain v_1 (because, by Prop. 2, the only elements of E of weight ω are the multiples of v, and v_1 is not a multiple of v). It is therefore distinct from V_1, and since V_1 is irreducible, one has $N_2 = 0$. Thus $f_2 \colon E \to V_2$ is an *isomorphism*. Similarly, one proves that E is isomorphic to V_1; therefore V_1 and V_2 are isomorphic.

Remark. One can give examples of irreducible $\mathfrak{g}$-modules which have no highest weight (in other words, which do not contain a primitive element); these modules are necessarily infinite dimensional (cf. Sec. 4).

Theorem 2. *For each $\omega \in \mathfrak{h}^*$, there is an irreducible $\mathfrak{g}$-module with highest weight equal to ω.*

(Theorem 1 shows that such a module is unique up to isomorphism.)

(i) Our first step will be to construct a $\mathfrak{g}$-module V_ω, containing a primitive element v of weight ω, and generated by v.

First let L_ω be a one-dimensional $\mathfrak{b}$-module, having as basis an element v such that

$$Hv = \omega(H) \quad \text{if } H \in \mathfrak{h}, \, Xv = 0 \quad \text{if } X \in \mathfrak{n}.$$

We can view L_ω as a $U\mathfrak{b}$-module, where $U\mathfrak{b}$ denotes the universal enveloping algebra of $\mathfrak{b}$. By taking a tensor product with $U\mathfrak{g}$, we obtain from this a $U\mathfrak{g}$-module

$$V_\omega = U\mathfrak{g} \otimes_{U\mathfrak{b}} L_\omega.$$

It is clear that the module V_ω is generated by the element $1 \otimes v$ (which we shall write simply as v); this element is nonzero since, by the Birkhoff–Witt Theorem, $U\mathfrak{g}$ is a free $U\mathfrak{b}$-module having a basis containing the unit element 1. Moreover, the formulae written above clearly show that v is primitive of weight ω.

(In fact, V_ω has as a *basis* the family of elements $Y_{\beta_1}^{m_1} \cdots Y_{\beta_k}^{m_k} v$, but we do not need this .)

(ii) Let V_ω be the module constructed above, and let us put

$$V_\omega^- = \sum_{\pi \neq \omega} (V_\omega)^\pi.$$

If V' is a $\mathfrak{g}$-submodule of V_ω distinct from V_ω, then $V \subset V_\omega^-$. Indeed since V' is stable under $\mathfrak{h}$, one has $V' = \sum V'^\pi$, and if V'^ω were nonzero, it would contain v and one would have $V' = V_\omega$. One therefore has $V' = \sum_{\pi \neq \omega} V'^\pi$, that is, $V' \subset V_\omega^-$. This being so, let N_ω be the $\mathfrak{g}$-submodule of V_ω generated by all the $\mathfrak{g}$-submodules of V_ω distinct from V_ω. By the above, one has $N_\omega \subset V_\omega^-$, so that $N_\omega \neq V_\omega$. The quotient module $E_\omega = V_\omega/N_\omega$ is obviously irreducible with highest weight ω.

Remarks. (1) Depending on the given linear form ω, it can happen that $N_\omega = 0$ or $N_\omega \neq 0$; both cases arise even for $\mathfrak{g} = \mathfrak{sl}_2$.

(2) Theorems 1 and 2 give a bijection between the elements ω of $\mathfrak{h}^*$ and the classes of irreducible $\mathfrak{g}$-modules with a highest weight.

4. Finite-Dimensional Modules

Proposition 3. *Let V be a finite-dimensional $\mathfrak{g}$-module. Then one has*

(a) $V = \sum V^\pi$.
(b) *If π is a weight of V, $\pi(H_\alpha)$ is an integer for all $\alpha \in R$.*
(c) *If $V \neq 0$, V contains a primitive element.*
(d) *If V is generated by a primitive element, V is irreducible.*

By Theorem 3 of Chap. III, the elements of $\mathfrak{h}$ are semisimple; the endomorphisms of V which they define are therefore diagonalizable (Chap. II, Theorem 7). Since they commute with each other, they can be diagonalized simultaneously, giving (a). Statement (c) follows from Lie's theorem (Chap. I, Theorem 3) applied to the solvable algebra $\mathfrak{b}$. Statement (d) follows from Prop. 2 (4), combined with the complete reducibility theorem (Chap. II, Theorem 8).

Finally, if $\alpha \in R^+$, V can be viewed as a module over the Lie algebra $\mathfrak{s}_\alpha$ generated by X_α, Y_α, H_α (cf. Chap. VI). By applying Theorem 4 of Chap. IV to this module, one sees that the eigenvalues of H_α on V belong to $\mathbf{Z}$. Since these eigenvalues are none other than the values $\pi(H_\alpha)$, one gets (b).

Corollary. *Every finite-dimensional irreducible $\mathfrak{g}$-module has a highest weight.*

This follows from (c).

In view of Theorems 1 and 2, it only remains to characterize the elements $\omega \in \mathfrak{h}^*$ which are highest weights of finite-dimensional irreducible modules.

Theorem 3. *Let $\omega \in \mathfrak{h}^*$ and let E_ω be an irreducible $\mathfrak{g}$-module having ω as highest weight. For E_ω to be finite dimensional, it is necessary and sufficient that one has*

$$(*) \quad \text{For all } \alpha \in R^+, \omega(H_\alpha) \text{ is an integer } \geqslant 0.$$

(Since the simple inverse roots H_i form a base for the inverse roots H_α, it is sufficient that the values $\omega(H_i)$ be integers $\geqslant 0$.)

The *necessity* of condition $(*)$ follows from the fact that, if v is a primitive element of E_ω for $\mathfrak{g}$, it is also a primitive element for the subalgebra $\mathfrak{s}_\alpha$ generated by X_α, Y_α, H_α. By Corollary 2 to Theorem 1 in Chap. IV, $\omega(H_\alpha)$ must therefore be an integer $\geqslant 0$.

Now let us show that condition $(*)$ is sufficient. Let v be a primitive element of E_ω, and let i be an integer between 1 and n. Let us put

$$m_i = \omega(H_i) \quad \text{and} \quad v_i = Y_i^{m_i+1} v.$$

If $j \neq i$, X_j and Y_i commute. One then has

$$X_j v_i = Y_i^{m_i+1} X_j v = 0.$$

Moreover, Theorem 1 of Chap. IV, applied to the subalgebra $\mathfrak{s}_i$ generated by X_i, Y_i, H_i, shows that $X_i v_i = 0$. If v_i were nonzero, it would then be a *primitive* element of E_ω, of weight $\omega - (m_i + 1)\alpha_i$, contradicting Theorem 1. This proves that $v_i = 0$. Theorem 1 of Chap. IV now shows that the vector subspace F_i of E_ω spanned by the elements $Y_i^p v$, $0 \leqslant p \leqslant m_i$, is a *finite-dimensional* $\mathfrak{s}_i$-*submodule of* E_ω.

Now let T_i be the set of finite-dimensional $\mathfrak{s}_i$-submodules of E_ω, and E_i' their sum. If $F \in T_i$, one checks easily that $\mathfrak{g} \cdot F \in T_i$; it follows that E_1' is a $\mathfrak{g}$-submodule of E_ω. Since E_ω is irreducible and E_i' nonzero (it contains F_i), we have $E_i' = E_\omega$. Thus we have proved that E_ω *is a sum of finite-dimensional* $\mathfrak{s}_i$-*submodules*.

Let P_ω be the set of weights of E_ω. We shall show that P_ω is *invariant under the symmetry* s_i associated with the root α_i. To see this, let $\pi \in P_\omega$, and let y be a nonzero element of E_ω^π. By Theorem 1, $p_i = \pi(H_i)$ is an integer. Let us put

$$x = Y_i^{p_i} y \quad \text{if } p_i \geqslant 0, \text{ and } x = X_i^{-p_i} y \quad \text{if } p_i \leqslant 0.$$

By Theorem 4 of Chap. IV, applied to $\mathfrak{s}_i$ and to a finite-dimensional $\mathfrak{s}_i$-submodule of E_ω containing y, one has $x \neq 0$. Since the weight of x is equal to

$$\pi - p_i \alpha_i = \pi - \omega(H_i)\alpha_i = s_i(\pi),$$

this shows that $s_i(\pi)$ is a weight of E_ω, and P_ω is indeed invariant under s_i.

Now let us prove that P_ω is *finite*. If $\pi \in P_\omega$, Theorem 1 shows that π can be written as

$$\pi = \omega - \sum p_i \alpha_i,$$

where the coefficients p_i are integers $\geqslant 0$. All that remains is to *bound* these coefficients. Now, because $-S$ is a base for R, there is an element w of the Weyl group of R sending S to $-S$, and this element is a product of the symmetries s_i (cf. Sec. V.10). It follows that $w(\pi)$ also belongs to P_ω, and can therefore be written as

$$w(\pi) = \omega - \sum q_i \alpha_i \quad \text{with } q_i \geqslant 0.$$

Applying w^{-1} to this formula, one finds

$$\pi = w^{-1}(\omega) + \sum r_i \alpha_i \quad \text{with } r_i \geqslant 0.$$

One concludes from this that $p_i + r_i$ is equal to the coefficient c_i of α_i in $\omega - w^{-1}(\omega)$; thus $p_i \leqslant c_i$, and the coefficients p_i are indeed bounded.

Thus, *there are only finitely many weights of* E_ω. Since each of them has finite multiplicity (Theorem 1), and since E_ω is the sum of the corresponding eigen-subspaces, E_ω is finite dimensional, as required.

Remarks. (1) In the course of this proof we have seen that the set P_ω of weights of E_ω is *invariant under the Weyl group W*. In fact, if $\pi \in P_\omega$ and $w \in W$, the weights π and $w(\pi)$ have the *same multiplicity*. For it is sufficient to see this when $w = s_i$, and in this case one easily checks that the element

$$\theta = e^{X_i} e^{-Y_i} e^{X_i}$$

sends the eigen-subspace corresponding to π to that corresponding to $s_i(\pi)$ (cf. Sec. IV.5, Remark 1).

(2) Let (ω_i) be the basis of $\mathfrak{h}^*$ dual to the basis (H_i):

$$\omega_i(H_i) = 1, \quad \omega_i(H_j) = 0 \qquad \text{if } i \neq j.$$

The ω_i are called the *fundamental weights* of the root system R (with respect to the chosen base S). Condition (∗) of Theorem 3 means that the linear form ω is a *linear combination of the weights ω_i, the coefficients being integers* ≥ 0.

The irreducible modules having the weights ω_i as highest weights are called the *fundamental modules* (or *fundamental representations*) of g.

5. An Application to the Weyl Group

Proposition 4. *The Weyl group W acts simply transitively on the set of bases of R.*

We know (Sec. V.10) that it acts transitively. Hence it is sufficient to prove that, if $w(S) = S$, with $w \in W$, then $w = 1$. Let P be the set of fundamental weights. We have $w(P) = P$. If $\omega \in P$, we know that $w(\omega)$ is a weight of the fundamental module E_ω with highest weight ω. By Theorem 1, it follows that $\omega - w(\omega)$ is a linear combination of the simple roots α_i, with coefficients ≥ 0. This applies to every $\omega \in P$. But on the other hand, we have

$$\sum_{\omega \in P} (\omega - w(\omega)) = \sum_{\omega \in P} \omega - \sum_{\omega \in P} \omega = 0.$$

This is impossible unless *each* of the summands $\omega - w(\omega)$ is zero. Since P is a basis for $\mathfrak{h}^*$, this indeed forces $w = 1$, as required.

6. Example: $\mathfrak{sl}_{n+1}$

Let g be the algebra $\mathfrak{sl}_{n+1}$ of square matrices of order $n + 1$ and trace zero. We take $\mathfrak{h}$ to be the subalgebra consisting of the diagonal matrices $H = (\lambda_1, \ldots, \lambda_{n+1})$, with $\sum \lambda_i = 0$. The roots are the linear forms $\alpha_{i,j}$, $i \neq j$, given by

$$\alpha_{i,j}(H) = \lambda_i - \lambda_j.$$

For a base, we take the roots $\alpha_i = \alpha_{i,i+1}$, $1 \leqslant i \leqslant n$. The element $H_i \in \mathfrak{h}$ corresponding to α_i has components $\lambda_i = 1$, $\lambda_{i+1} = -1$, $\lambda_j = 0$ if $j \neq i, i+1$.

The fundamental weights ω_i are given by

$$\omega_i(H) = \lambda_1 + \cdots + \lambda_i.$$

The fundamental weight ω_1 is the highest weight of the natural representation of $\mathfrak{sl}_{n+1}$ on the vector space $E = \mathbf{C}^{n+1}$. More generally, ω_i is the highest weight of the *i-th exterior power* of E.

(In fact, all the finite-dimensional irreducible representations of $\mathfrak{sl}_{n+1}$ can be obtained by decomposing the tensor powers of E; for more details, see H. Weyl, *The Classical Groups*.)

7. Characters

Let P be the subgroup of $\mathfrak{h}^*$ consisting of the elements π such that $\pi(H_\alpha) \in \mathbf{Z}$ for all $\alpha \in R$ (or equivalently, for all $\alpha \in S$). The group P is a free abelian group, having a basis consisting of the fundamental weights $\omega_1, \ldots, \omega_n$.

We will denote by A the group-algebra $\mathbf{Z}[P]$ of the group P with coefficients in $\mathbf{Z}$. By definition, A has a basis $(e^\pi)_{\pi \in P}$ such that $e^\pi \cdot e^{\pi'} = e^{\pi+\pi'}$.

Definition. *Let V be a finite-dimensional $\mathfrak{g}$-module. For each weight π of V, let $m_\pi = \dim V^\pi$ be the multiplicity of π in V. The element*

$$\mathrm{ch}(V) = \sum m_\pi e^\pi$$

of the algebra A is called the character of V.

(This definition is legitimate, since each weight π of V belongs to P by Prop. 3.)

Proposition 5. (a) $\mathrm{ch}(V)$ *is invariant under the Weyl group W.*

(b) *One has*

$$\mathrm{ch}(V \oplus V') = \mathrm{ch}(V) + \mathrm{ch}(V')$$

$$\mathrm{ch}(V \otimes V') = \mathrm{ch}(V) \cdot \mathrm{ch}(V').$$

(c) *Two finite-dimensional $\mathfrak{g}$-modules V and V' are isomorphic if and only if* $\mathrm{ch}(V) = \mathrm{ch}(V')$,

Statement (a) expresses the fact that two weights which are equivalent under W have the same multiplicity (cf. Sec. 4, Remark 1). The two formulae in (b) are obvious. Let us now prove (c). We must show that $\mathrm{ch}(V) = \mathrm{ch}(V')$ implies that V and V' are isomorphic. We argue by induction on $\dim V$. If $\dim V = 0$, then $\mathrm{ch}(V) = 0$, so that $V' = 0$. Otherwise, let P_V be the set of weights of V (which is the same as that for V' since the characters of V and V' are equal).

We have $P_V \ne \varnothing$, and since P_V is finite, one can find an element $\omega \in P_V$ such that $\omega + \alpha_i$ does not belong to P_V for any i. If v is a nonzero element of V^ω, it is clear that v is primitive. By Prop. 3, the submodule V_1 of V generated by v is irreducible and has highest weight ω. By the complete reducibility theorem, one has $V = V_1 \oplus V_2$, where V_2 is a submodule of V. The same argument, applied to V', shows that $V' = V_1' \oplus V_2'$, where V_1' is irreducible with highest weight ω. Since V_1 and V_1' have the same highest weight, they are isomorphic (Theorem 1), so that $\mathrm{ch}(V_1) = \mathrm{ch}(V_1')$. Using (b), we now see that $\mathrm{ch}(V_2) = \mathrm{ch}(V_2')$, and the induction hypothesis shows that V_2 is isomorphic to V_2'; thus V and V' are isomorphic, as required.

Let A^W be the subalgebra of $A = \mathbf{Z}[P]$ consisting of the elements invariant under the Weyl group W. By (a), every character belongs to A^W; conversely, every element of A^W is the difference of two characters. This is a consequence of the following more precise proposition, which we shall not prove (for a proof, see Bourbaki, Chap. 8, Sec. 7, No. 7).

Proposition* 6. *Let $T_i = \mathrm{ch}(E_{\omega_i})$ be the character of the i-th fundamental module of $\mathfrak{g}$ (cf. Sec. 4, Remark 2). The elements T_i, $1 \leqslant i \leqslant n$, are algebraically independent, and generate the algebra A^W.*

One can therefore identify A^W with the polynomial algebra

$$\mathbf{Z}[T_1, \ldots, T_n].$$

Corollary. *The map defined by* ch *induces an isomorphism from the "Grothendieck group" of finite-dimensional $\mathfrak{g}$-modules onto the algebra A^W.*

This follows from Propositions 5 and 6.

8. H. Weyl's Formula

This formula allows one to calculate the character of an irreducible $\mathfrak{g}$-module as a function of its highest weight.

Before giving it, let us introduce some notation:

(i) If $w \in W$, let $\varepsilon(w)$ denote the determinant of w. This is $+1$ if w is the product of an *even* number of symmetries s_α, and -1 otherwise.

(ii) We put $\rho = \frac{1}{2}\sum_{\alpha > 0} \alpha$; one can show that $\rho(H_i) = 1$ for all i, so that $\rho \in P$.

(iii) We put

$$D = \prod_{\alpha > 0} (e^{\alpha/2} - e^{-\alpha/2}),$$

the product being evaluated in the algebra $\mathbf{Z}[\frac{1}{2}P]$. In fact, we have $D \in \mathbf{Z}[P]$, since one can show that

$$D = \sum_{w \in W} \varepsilon(w) e^{w(\rho)}.$$

Theorem* 4. *Let E be a finite-dimensional irreducible g-module, and ω its highest weight. One has*

$$\text{ch}(E) = \frac{1}{D} \cdot \sum_{w \in W} \varepsilon(w) e^{w(\omega + \rho)}.$$

The original proof of this theorem (Weyl, 1926) used the theory of compact groups (cf. Séminaire S. Lie, exposé 21). A "purely algebraic" (but less natural) proof was found in 1954 by Freudenthal; it is reproduced in Jacobson's book (see also Bourbaki, Chap. 8, Sec. 9).

Corollary 1. *The dimension of E is given by the formula*

$$\dim E = \prod_{\alpha > 0} \frac{\langle \omega + \rho, H_\alpha \rangle}{\langle \rho, H_\alpha \rangle} = \prod_{\alpha > 0} \frac{(\omega + \rho, \alpha)}{(\rho, \alpha)}.$$

One deduces this from the theorem by computing the sum of the coefficients of ch(E) (cf. Bourbaki, Chap. 8, Sec. 9).

Corollary 2. *Let V be a finite-dimensional g-module, and let n(V, ω) be the multiplicity with which E appears in a decomposition of V as a direct sum of irreducible modules. Then n(V, ω) is equal to the coefficient of $e^{\omega + \rho}$ in the product $D \cdot \text{ch}(V)$.*

This is a simple consequence of the theorem.

EXAMPLE. For $g = \mathfrak{sl}_2$, there is a unique positive root α equal to 2ρ. The group P consists of the integer multiples of ρ. A highest weight ω can be written as $\omega = m\rho$, with $m \geqslant 0$. Weyl's formula gives

$$\text{ch}(E) = \frac{e^{(m+1)\rho} - e^{-(m+1)\rho}}{e^\rho - e^{-\rho}} = e^{m\rho} + e^{(m-2)\rho} + \cdots + e^{-m\rho},$$

which is indeed consistent with the results in Chap. IV.

CHAPTER VIII

Complex Groups and Compact Groups

This chapter contains no proofs. All the Lie groups considered (except in Sec. 7) are *complex* groups.

1. Cartan Subgroups

From now on, G denotes a connected Lie group whose Lie algebra $\mathfrak{g}$ is semisimple. Such a group is called a *complex semisimple group*.

Let $\mathfrak{h}$ be a Cartan subalgebra of $\mathfrak{g}$, and let H be the Lie subgroup of G corresponding to $\mathfrak{h}$. The conjugates of H are called the *Cartan subgroups* of G.

Theorem 1. (a) H *is a closed group subvariety of* G. (b) H *is a group of multiplicative type* (i.e. isomorphic to a product of groups $\mathbf{C}^*$).

Let us describe the structure of H more precisely. Let R be the root system of $\mathfrak{g}$ with respect to $\mathfrak{h}$, let $R^* \subset \mathfrak{h}$ be the inverse system, let Γ be the subgroup of $\mathfrak{h}$ generated by the elements H_α of R^*, and let Γ_1 be the subgroup of $\mathfrak{h}$ consisting of those $x \in \mathfrak{h}$ such that $\alpha(x) \in \mathbf{Z}$ for all $\alpha \in R$.
One has

$$\Gamma \subset \Gamma_1 \subset \mathfrak{h}.$$

Furthermore, let

$$e: \mathfrak{h} \to H$$

be the map $x \mapsto \exp(2i\pi x)$. This is a homomorphism, since $\mathfrak{h}$ is abelian.

Theorem 2. (a) *The homomorphism* $e: \mathfrak{h} \to H$ *is surjective, and its kernel* $\Gamma(G)$ *lies between* Γ *and* Γ_1.

(b) *e defines an isomorphism from* $\Gamma_1/\Gamma(G)$ *onto the center of G.*

(c) *The canonical map* $\pi_1(H) \to \pi_1(G)$ *is surjective. It induces an isomorphism from the the quotient* $\Gamma(G)/\Gamma$ *onto* $\pi_1(G)$. (One identifies $\pi_1(H)$ with $\Gamma(G)$ by means of *e*.)

Statement (a) has a converse.

Theorem 3. *For each subgroup M of* $\mathfrak{h}$ *such that* $\Gamma \subset M \subset \Gamma_1$, *there is a complex semisimple group G, with Lie algebra* $\mathfrak{g}$, *such that* $\Gamma(G) = M$. *This group is unique (up to a unique isomorphism).*

In particular, G is an *adjoint group* (trivial center) if and only if $\Gamma(G) = \Gamma_1$; G is *simply connected* $(\pi_1(G) = 0)$ if and only if $\Gamma(G) = \Gamma$.

Since the group Γ_1/Γ is finite, we see that there are only *a finite number* of complex semisimple groups with a given Lie algebra—up to isomorphism, of course. They are obtained by taking the coverings of the adjoint group.

EXAMPLES. (1) If $\mathfrak{g} = \mathfrak{sl}_2$, Γ is the set $\mathbf{Z}\alpha$ of multiples of a root α, and $\Gamma_1 = \frac{1}{2}\mathbf{Z}\alpha$. Hence there are *two* groups: the simply connected group SL_2 and the adjoint group $PGL_2 = SL_2/\{\pm 1\}$.

(2) Here is the structure of Γ_1/Γ for each of the types of simple Lie algebras:

type $A_n, n \geqslant 1$:	$\Gamma_1/\Gamma = \mathbf{Z}/(n+1)\mathbf{Z}$
type D_n, even $n \geqslant 4$:	$\Gamma_1/\Gamma = \mathbf{Z}/2\mathbf{Z} \times \mathbf{Z}/2\mathbf{Z}$
type D_n, odd $n \geqslant 5$:	$\Gamma_1/\Gamma = \mathbf{Z}/4\mathbf{Z}$
type E_6:	$\Gamma_1/\Gamma = \mathbf{Z}/3\mathbf{Z}$
types $B_n, C_n, n \geqslant 2$, and E_7:	$\Gamma_1/\Gamma = \mathbf{Z}/2\mathbf{Z}$
types G_2, F_4, E_8:	$\Gamma_1/\Gamma = 0$.

2. Characters

One defines a *character* of the Cartan group H to be any (complex analytic) homomorphism $\chi: H \to \mathbf{C}^*$. The group $X(H)$ of characters of H is a free abelian group of rank equal to dim H. Knowledge of $X(H)$ is *equivalent* to knowledge of H, by virtue of the formula

$$H = \mathrm{Hom}(X(H), \mathbf{C}^*).$$

If $\chi \in X(\mathbf{H})$, the tangent mapping at χ is a homomorphism from $\mathfrak{h}$ to $\mathbf{C}$, that is, an element of $\mathfrak{h}^*$. Thus we get an injection from $X(H)$ into $\mathfrak{h}^*$, which allows us to *identify* $X(H)$ *with a subset of* $\mathfrak{h}^*$. With this convention, we have:

Theorem 4. (a) *Let P_1 (resp. P) be the subgroup of $\mathfrak{h}^*$ generated by the roots (resp. by the highest weights). One has $P_1 \subset X(H) \subset P$.*

(b) *Conversely, every subgroup M^* lying between P_1 and P is an $X(H)$ for a suitable choice of G; the corresponding group G is then defined up to a unique isomorphism.*

This assertion is simply a reformulation of Theorem 3, if one notices that P, P_1, and $X(H)$ are the duals of Γ, Γ_1, and $\Gamma(G)$.

3. Relations with Representations

Let $\rho\colon \mathfrak{g} \to \operatorname{End}(E)$ be a finite-dimensional linear representation of $\mathbf{g}$.

Theorem 5. *For ρ to correspond to a linear representation*

$$\tilde{\rho}\colon G \to \operatorname{GL}(E)$$

of the group G, it is necessary and sufficient that the weights of E belong to the subgroup $X(H)$ defined in Sec. 2.

Remarks. (1) This condition is always satisfied if G is simply connected.

(2) If E is irreducible, with highest weight ω (with respect to a base S of R), it is sufficient that ω should belong to $X(H)$. Indeed, we know (Chap. VII) that every weight of E has the form $\omega - \gamma$, with $\gamma \in P$.

EXAMPLE. Let us take G to be the adjoint group of $\mathfrak{sl}_2$, that is, $\operatorname{PGL}_2 = \operatorname{SL}_2/\{\pm 1\}$. By the above, the irreducible representations of G correspond to the irreducible representations of $\mathfrak{sl}_2$ for which the highest weight is an *integer* multiple of the positive root α; these are the representations W_{2n} of Chap. IV.

4. Borel Subgroups

Let S be a base for R, and let $\mathfrak{g} = \mathfrak{n}^- \oplus \mathfrak{h} \oplus \mathfrak{n}$ be the corresponding decomposition of $\mathfrak{g}$ (cf. Sec. VI.6). Let $\mathfrak{b} = \mathfrak{h} \oplus \mathfrak{n}$.

Theorem 6. (a) *The exponential map defines an isomorphism from $\mathfrak{n}$ onto a closed group subvariety U of G.*

(b) *The Lie subgroup B corresponding to $\mathfrak{b}$ is closed; it is the semidirect product of H and U.*

(c) *Every connected solvable subgroup of G is conjugate to a subgroup of H.*

The groups conjugate to B are called the *Borel subgroups* of G.

Theorem 7. *The quotient G/B is a projective algebraic variety* (hence compact).

Moreover, one has a cellular decomposition of G/B, called the "*Bruhat decomposition*", which corresponds to the action of B on G/B.

More precisely, let W be the Weyl group of $\mathfrak{g}$ with respect to $\mathfrak{h}$. If $w \in W$, let us choose an element $n_w \in G$ such that $\mathrm{Int}(n_w)$ leaves $\mathfrak{h}$ invariant and induces w on $\mathfrak{h}$ (cf. Sec. VI.2); let $\bar{w}$ be the image of n_w in G/B.

Theorem 8. *G/B is the disjoint union of the orbits $B\bar{w}$, for $w \in W$.*

Corollary. *The map $w \mapsto Bn_w B$ is a bijection from W onto the set $B\backslash G/B$ of double cosets of G modulo B.*

Moreover, each $B\bar{w}$ is isomorphic to an affine space $\mathbf{C}^{n(w)}$.

5. Construction of Irreducible Representations from Borel Subgroups

Let us keep the notation of the preceding section. Let ω be a highest weight, that is, an element of P such that $\omega(H_\alpha) \geqslant 0$ for all $\alpha \in S$. Suppose that ω belongs to the subgroup $X(H)$ of P. Then by Sec. 3, there exists an irreducible representation E_ω of G, with highest weight ω. We shall describe it explicitly.

We have seen that ω may be viewed as a homomorphism from H to $\mathbf{C}^*$; let us extend this homomorphism to the whole of B by putting

$$\omega(u) = 1 \qquad \text{for all } u \in U.$$

Let V_ω be the set of holomorphic functions f on G which satisfy the identity

$$f(yb) = \omega(b)f(y) \qquad \text{if } b \in B, \, y \in G.$$

We make G act on V_ω by putting

$$(g \, f)(y) = f(g^{-1}y) \qquad \text{if } y, g \in G.$$

Theorem 9. *The representation V_ω is irreducible and finite dimensional. Its dual is isomorphic to E_ω.*

If $e \in E_\omega$ is a primitive element of weight ω, one defines a G-map

$$i: (E_\omega)^* \to V_\omega$$

by putting

$$i(\lambda)(y) = \langle \lambda, y \cdot e \rangle.$$

One can show that this is an isomorphism.

[To prove that V_ω is finite dimensional, one puts on V_ω the topology of convergence on compact subsets. Using the compactness of G/B, one sees that V_ω is a Banach space which is locally compact (by Montel's criterion) and hence finite dimensional.]

6. Relations with Algebraic Groups

Let G be a complex semisimple group.

Theorem 10. (a) *There is a complex algebraic group structure on G, and one only, which is compatible with its analytic group structure.*

(b) *If G' is a complex algebraic group, every analytic homomorphism from G to G' is algebraic.*

Thus we have a dictionary "algebraic $\Leftrightarrow$ analytic".

Remark. The algebraic structure on G can be described in several ways. The most "explicit" consists of choosing a faithful linear representation $\tilde{\rho}: G \to \mathrm{GL}(E)$, and showing that $\tilde{\rho}(G)$ is an algebraic subgroup of $\mathrm{GL}(E)$; hence $\tilde{\rho}$ induces an algebraic structure on G, which can be shown to be independent of the choice of $\tilde{\rho}$.

One can also give a direct definition of the *affine algebra* A_G of the algebraic structure on G, as follows:

$$f \in A_G \Leftrightarrow f \text{ is a holomorphic function on } G \text{ whose translates}$$
$$\text{span a finite-dimensional vector space.}$$

7. Relations with Compact Groups

(In this section, we are concerned with both *real* Lie groups and *complex* Lie groups. For the results stated here, see Chevalley, *Theory of Lie groups*, Chap. 6, Secs. 8–12.)

First we recall:

Theorem 11. *Let K be a real connected semisimple Lie group, with Lie algebra $\mathfrak{k}$. For K to be compact, it is necessary and sufficient that the Killing form of $\mathfrak{k}$ should be negative.*

Here now is the assertion which underlies Weyl's "unitarian trick" (cf. Corollary 2).

Theorem 12. *Let G be a complex semisimple Lie group, with Lie algebra $\mathfrak{g}$. Let K be a maximal compact subgroup of G, with Lie algebra $\mathfrak{k}$.*

(a) *The algebra* $\mathfrak{k}$ *is a real form of* $\mathfrak{g}$, *that is,* $\mathfrak{g} = \mathfrak{k} \oplus i\mathfrak{k}$.

(b) *The exponential map defines an isomorphism* (of real analytic variety structures) *from* $i\mathfrak{k}$ *onto a closed subvariety N of G.*

(c) *The map* $(k, n) \mapsto k \cdot n$ *is an isomorphism* (of real analytic varieties) *from* $K \times N$ *onto G.*

(d) *Every compact subgroup of G is contained in a conjugate of K.*

This theorem has the following consequences:

Corollary 1. *The homology groups and homotopy groups of G can be identified with those of K. In particular,* $\pi_1(G) = \pi_1(K)$.

This follows from (b) and (c).

Corollary 2. *If G' is a complex Lie group, the restriction map*

$$r: \operatorname{Hom}_{\mathbf{C}}(G, G') \to \operatorname{Hom}_{\mathbf{R}}(K, G')$$

is bijective.

This follows from (a) and the fact that $\pi_1(G) = \pi_1(K)$.

Corollary 3. *The group G is determined up to a unique isomorphism by K.*

This follows from Corollary 2.

Conversely:

Theorem 13. *Let K be a compact, connected Lie group, with a semisimple Lie algebra. There exists a complex semisimple Lie group G which contains K as a maximal compact subgroup.*

(In view of the preceding result, this group is unique, up to a unique isomorphism; it is called the *complexification* of K.)

The construction of the affine algebra A_G of G is easy: its elements are the continuous functions f on K whose translates span a finite-dimensional vector space. Notice that this algebra has a canonical real structure; it therefore defines *an algebraic group L over* **R**. The set of real points of L is K, its set of complex points is G.

Bibliography

N. Bourbaki. *Groupes et Algèbres de Lie*, Chap. 1–9. Paris, Hermann-Masson, 1960–1982.

E. Cartan. *Oeuvres Complètes, I-1*. Paris, Gauthier-Villars, 1952.

R. W. Carter. Simple groups and simple Lie algebras. *J. London Math. Soc.*, **40** (1965) 193–240.

R. W. Carter. *Simple Groups of Lie Type*. Wiley, London, 1972.

C. Chevalley. *Theory of Lie Groups*. Princeton Univ. Press, 1946.

C. Chevalley. Sur certains groupes simples. Tôhoku Math. J., **7** (1955), 14–66.

C. Chevalley. Classification des Groupes de Lie Algébriques. Séminaire, Paris, 1958.

M. Demazure and A. Grothendieck. Schémas en groupes. Séminaire IHES, 1963–1964. Lecture Notes in Mathematics, Nos. 151, 152, 153, Springer-Verlag, Heidelberg, 1970.

W. Fulton and J. Harris. *Representation Theory*. GTM 129, Springer-Verlag, 1991.

S. Helgason. *Differential Geometry and Symmetric Spaces*. Academic Press, New York, 1962.

G. Hochschild. *The Structure of Lie Groups*. Holden-Day, San Francisco, 1965.

J. E. Humphreys. *Introduction to Lie Algebras and Representation Theory*. Springer-Verlag, New York, 1972.

N. Jacobson. *Lie Algebras*. Interscience Tracts No. 10, John Wiley and Sons, New York, 1962.

Séminaire Sophus LIE. *Théorie des Algèbres de Lie – Topologie des Groupes de Lie*. Paris, 1955.

J-P. Serre. *Lie Algebras and Lie Groups*. Benjamin, New York, 1965; Lect. Notes in Math. 1500, Springer-Verlag, 1992.

H. Weyl. *Gesammelte Abhandlungen*. 4 vols., Springer-Verlag, Heidelberg, 1968.

Index